T0137124

Terrorism, Security, and Computation

Series Editor
V.S. Subrahmanian

More information about this series at http://www.springer.com/series/11955

Kishan G. Mehrotra • Chilukuri K. Mohan
HuaMing Huang

Anomaly Detection
Principles and Algorithms

 Springer

Kishan G. Mehrotra
Department of Electrical Engineering
and Computer Science
Syracuse University
Syracuse, NY, USA

Chilukuri K. Mohan
Department of Electrical Engineering
and Computer Science
Syracuse University
Syracuse, NY, USA

HuaMing Huang
Department of Electrical Engineering
and Computer Science
Syracuse University
Syracuse, NY, USA

ISSN 2197-8778   ISSN 2197-8786   (electronic)
Terrorism, Security, and Computation
ISBN 978-3-319-88445-5   ISBN 978-3-319-67526-8   (eBook)
https://doi.org/10.1007/978-3-319-67526-8

Printed on acid-free paper

This Springer imprint is published by Springer Nature
The registered company is Springer International Publishing AG
The registered company address is: Gewerbestrasse 11, 6330 Cham, Switzerland

To Rama, Sudha, Lin, and Stephen

Preface

This book addresses the research area of anomaly detection algorithms, beginning with conceptual foundations, and then delving into algorithmic details, with examples drawn from several practical application areas. In one form or the other, researchers have been studying anomaly detection for over a hundred years, mostly with isolated results that are unknown to those who pursue other fields of enquiry.

Many new advances have been made in the past decades, partly thanks to the rapid pace of advancements in pattern recognition and data mining, along with the availability of powerful computer hardware that facilitates addressing problems that were previously considered intractable. We still search for needles in haystacks, but our task is far more likely to be achievable than in previous decades.

Anomalies arise in numerous fields of study, including medicine, finance, cyber-security, sociology, and astronomy. Indeed, anomalies are believed to drive scientific innovation, requiring us to develop an understanding of observations that are not well explained by prior hypotheses and accepted beliefs. In some areas, anomalies indicate problematic behavior, e.g., unusual purchases that indicate credit card fraud. In some other areas, they may be indicators of positive outcomes, e.g., unexpectedly higher sales within a retail organization. In other cases, they may merely indicate ill-understood phenomena or unknown objects or processes, triggering the exploration of new ideas that enrich the field of human enquiry.

Detection of anomalies is sometimes an art, sometimes a craft, and sometimes just a matter of being lucky, as those who study subatomic particle physics or astronomy will testify. Sometimes an anomaly detection methodology successful in one domain can also prove successful in a completely new area of study, and knowledge of the former can enable more rapid advances in the latter. It is hence important to have a firm grasp of the principles and algorithms of anomaly detection, and understand the scope of their applicability.

The first five chapters of this book comprise Part I, providing a gentle introduction to various concepts and applications of anomaly detection, and the flavors of various distance measures and perspectives about the anomaly detection problem, including anomalies seen as out-of-cluster points, and addressing the notion of anomalies in the context of time series, requiring specialized approaches. These chapters are expected to be comprehensible by individuals with a bachelor's degree in any technical field, preferably with prior exposure to the fields of probability and statistics.

The last four chapters comprise Part II, and build on Part I, focusing on algorithms of various kinds; the explanations are written to be understandable by students and professionals from various fields, although the implementation of some of the more elaborate algorithms will require significant programming capabilities. Fortunately, many of these algorithms have been implemented by others and readily available from various sources on the internet. The most well-known algorithms have been implemented in libraries available with many data mining tools, packages, and libraries, such as Weka and MATLAB.

Unique features of this book include a coverage of rank-based anomaly detection algorithms, recently developed by the authors, and ensemble algorithms that successfully combine ideas from many individual algorithms. In addition, considerable space is devoted to discussing anomaly detection problems that arise in the context of sequential data or time series, i.e., where understanding the data requires knowing the exact sequence in which different data points arise.

This book has been the product of several years of discussions and work on applied research projects by the authors, and some algorithmic solutions and perspectives have arisen as direct results of such exposure to difficult real-life problems, many of which involve very large amounts of data. We thank our collaborators for the past two decades for insights gained into such problems, as well as several colleagues and students at Syracuse University who have participated in discussions on these topics, with particular thanks to Yiou Xiao and Zhiruo Zhao for assisting with the preparation of material in the book. Finally, we thank our collaborators at Springer for encouraging this effort and demonstrating great patience while delaying the production schedule over the past three years.

Syracuse, NY, USA   Kishan G. Mehrotra
Syracuse, NY, USA   Chilukuri K. Mohan
Syracuse, NY, USA   HuaMing Huang

Contents

Part I Principles

1 Introduction 3
  1.1 What's an Anomaly? 4
  1.2 Cybersecurity 7
      1.2.1 Privacy 7
      1.2.2 Malware Detection 8
      1.2.3 Fraudulent Email 8
  1.3 Finance 9
      1.3.1 Credit Card Fraud 9
      1.3.2 Creditworthiness 10
      1.3.3 Bankruptcy Prediction 10
      1.3.4 Investing 10
  1.4 Healthcare 11
      1.4.1 Diagnosis 11
      1.4.2 Patient Monitoring 12
      1.4.3 Radiology 12
      1.4.4 Epidemiology 12
  1.5 Defense and Internal Security 12
      1.5.1 Personnel Behaviors 13
      1.5.2 Battlefield Behaviors 13
      1.5.3 Unconventional Attacks 13
  1.6 Consumer Home Safety 14
      1.6.1 Detecting Occurrence of Falls and Other Problems 14
      1.6.2 Home Perimeter Safety 15
      1.6.3 Indoor Pollution Monitoring 15
  1.7 Manufacturing and Industry 16
      1.7.1 Quality Control 16
      1.7.2 Retail Sales 16

       1.7.3   Inventory Management 17

       1.7.4   Customer Behavior 17

       1.7.5   Employee Behavior 17

  1.8   Science 18

  1.9   Conclusion 19

2 Anomaly Detection 21

  2.1   Anomalies 21

       2.1.1   Metrics for Measurement 23

       2.1.2   Old Problems vs. New Problems 24

       2.1.3   What Kind of Data? 24

       2.1.4   What's a Norm? 25

  2.2   Outliers in One-Dimensional Data 26

  2.3   Outliers in Multidimensional Data 28

  2.4   Anomaly Detection Approaches 29

  2.5   Evaluation Criteria 30

  2.6   Conclusion 32

3 Distance-Based Anomaly Detection Approaches 33

  3.1   Introduction 33

  3.2   Similarity Measures 34

  3.3   Distance-Based Approaches 36

       3.3.1   Distance to All Points 36

       3.3.2   Distance to Nearest Neighbor 37

       3.3.3   Average Distance to k Nearest Neighbors 37

       3.3.4   Median Distance to k Nearest Neighbors 38

  3.4   Conclusion 39

4 Clustering-Based Anomaly Detection Approaches 41

  4.1   Identifying Clusters 41

       4.1.1   Nearest Neighbor Clustering 42

       4.1.2 k-Means Clustering 43

       4.1.3   Fuzzy Clustering 45

       4.1.4   Agglomerative Clustering 46

       4.1.5   Density-Based Agglomerative Clustering 47

       4.1.6   Divisive Clustering 48

  4.2   Anomaly Detection Using Clusters 49

       4.2.1   Cluster Membership or Size 49

       4.2.2   Proximity to Other Points 50

       4.2.3   Proximity to Nearest Neighbor 51

       4.2.4   Boundary Distance 51

       4.2.5   When Cluster Sizes Differ 53

       4.2.6   Distances from Multiple Points 54

  4.3   Conclusion 55

5   Model-Based Anomaly Detection Approaches 57

   5.1   Models of Relationships Between Variables 57

        5.1.1   Model Parameter Space Approach 58

        5.1.2   Data Space Approach 59

   5.2   Distribution Models 63

        5.2.1   Parametric Distribution Estimation 63

        5.2.2   Regression Models 64

   5.3   Models of Time-Varying Processes 67

        5.3.1   Markov Models 70

        5.3.2   Time Series Models 72

   5.4   Anomaly Detection in Time Series 78

        5.4.1   Anomaly Within a Single Time Series 79

        5.4.2   Anomaly Detection Among Multiple Time Series 84

   5.5   Learning Algorithms Used to Derive Models from Data 91

        5.5.1   Regularization 92

   5.6   Conclusion 93

Part II   Algorithms

6   Distance and Density Based Approaches 97

   6.1   Distance from the Rest of the Data 97

        6.1.1   Distance Based-Outlier Approach 100

   6.2   Local Correlation Integral (LOCI) Algorithm 102

        6.2.1   Resolution-Based Outlier Detection 104

   6.3   Nearest Neighbor Approach 105

   6.4   Density Based Approaches 107

        6.4.1   Mixture Density Estimation 109

        6.4.2   Local Outlier Factor (LOF) Algorithm 110

        6.4.3   Connectivity-Based Outlier Factor (COF) Approach 112

        6.4.4   INFLuential Measure of Outlierness by Symmetric
                Relationship (INFLO) 114

   6.5   Performance Comparisons 116

   6.6   Conclusions 117

7   Rank Based Approaches 119

   7.1   Rank-Based Detection Algorithm (RBDA) 121

        7.1.1   Why Does RBDA Work? 122

   7.2   Anomaly Detection Algorithms Based on Clustering and
        Weighted Ranks 124

        7.2.1   NC-Clustering 125

        7.2.2   Density and Rank Based Detection Algorithms 126

   7.3   New Algorithms Based on Distance and Cluster Density 127

   7.4   Results 130

        7.4.1   RBDA Versus the Kernel Based Density Estimation
                Algorithm 130

7.4.2   Comparison of RBDA and Its Extensions with LOF,
         COF, and INFLO 131
7.4.3   Comparison for KDD99 and Packed Executables
         Datasets 133
7.5   Conclusions... 134

8   Ensemble Methods.. 135
8.1   Independent Ensemble Methods..................................... 135
8.2   Sequential Application of Algorithms............................... 139
8.3   Ensemble Anomaly Detection with Adaptive Sampling 140
      8.3.1   AdaBoost.. 141
      8.3.2   Adaptive Sampling 142
      8.3.3   Minimum Margin Approach 142
8.4   Weighted Adaptive Sampling 143
      8.4.1   Weighted Adaptive Sampling Algorithm 147
      8.4.2   Comparative Results....................................... 147
      8.4.3   Dataset Description.. 148
      8.4.4   Performance Comparisons 148
      8.4.5   Effect of Model Parameters 151
8.5   Conclusion.. 152

9   Algorithms for Time Series Data................................... 153
9.1   Problem Definition 154
9.2   Identification of an Anomalous Time Series 157
      9.2.1   Algorithm Categories...................................... 158
      9.2.2   Distances and Transformations 159
9.3   Abnormal Subsequence Detection................................... 167
9.4   Outlier Detection Based on Multiple Measures...................... 169
      9.4.1   Measure Selection 169
      9.4.2   Identification of Anomalous Series 172
9.5   Online Anomaly Detection for Time Series......................... 172
      9.5.1   Online Updating of Distance Measures...................... 173
      9.5.2   Multiple Measure Based Abnormal Subsequence
               Detection Algorithm (MUASD) 176
      9.5.3   Finding Nearest Neighbor by Early Abandoning 178
      9.5.4   Finding Abnormal Subsequence Based on Ratio of
               Frequencies (SAXFR)...................................... 179
      9.5.5   MUASD Algorithm... 181
9.6   Experimental Results.. 182
      9.6.1   Detection of Anomalous Series in a Dataset 182
      9.6.2   Online Anomaly Detection.................................. 183
      9.6.3   Anomalous Subsequence Detection 186
      9.6.4   Computational Effort 188
9.7   Conclusion.. 188

Appendix A   Datasets for Evaluation 191
   A.1   A   Datasets for Evaluation 191
   A.2   Real Datasets 191
   A.3   KDD and PED 194

Appendix B   Datasets for Time Series Experiments 195
   B.1   Datasets 195
      B.1.1   Synthetic Datasets 195
      B.1.2   Brief Description of Datasets 195
      B.1.3   Datasets for Online Anomalous Time Series Detection 202
      B.1.4   Data Sets for Abnormal Subsequence Detection in a
         Single Series 203
      B.1.5   Results for Abnormal Subsequence Detection in a
         Single Series for Various Datasets 203

References 209

Index 215

List of Figures

Fig. 1.1   **Distribution of Intelligence Quotient (IQ) scores**—Mean of 100, Standard Deviation of 15 5

Fig. 1.2   **Monthly sales for a retailer**—December 2016 sales are anomalous with respect to prior year December sales.............. 5

Fig. 3.1   Dot plot of dataset $\mathscr{D} = \{1, 2, 3, 8, 20, 21\}$; *red dot* corresponds to anomalous observation 36

Fig. 3.2   Dot plot of dataset $\mathscr{D} = \{1, 3, 5, 7, 100, 101, 200, 202, 205, 208, 210, 212, 214\}$, *red dot* indicates anomalous observation 37

Fig. 3.3   Dot plot of dataset $\mathscr{D} = \{1, 3, 5, 7, 100, 101, 200, 202, 205, 208, 210, 212, 214\}$, *red dot* indicates anomalous observation 38

Fig. 4.1   Asymmetric clusters and the results of applying k-means algorithm with $k = 3$; clusters are identified by the colors of the points 44

Fig. 4.2   Point A is a core point because its neighbor contains Eps $= 6$ points; Point B is a border point because it belongs to A's neighborhood but its neighborhood does not contain Eps points; point C is a noise/anomalous point 47

Fig. 5.1   Feedforward Neural Network with a hidden layer.................. 66

Fig. 5.2   Three HMM models with conditional probability of H and T as indicated by *round arrows*; a *straight arrow* provides the probability of transition from one state to another. The leftmost figure is for a fair coin, the second is biased (in favor of Heads), and the third figure is symmetric but has low probability of going from one state to another 70

Fig. 5.3   Model parameter *Prob(Heads)* based on all preceding time points 71

Fig. 5.4   Result of periodically updating the parameter *Prob(Heads)* after 5 coin tosses 72

Fig. 5.5   Example monthly sales chart for a retail company 73

Fig. 5.6   Basis functions of Haar transformation, for data with four
            observations 77
Fig. 5.7   Basis functions of Haar transformation, for data with eight
            observations 78
Fig. 5.8   Daily closing prices for a company's stock in 2015 79
Fig. 5.9   Daily closing stock prices, with an anomaly near day 61 80
Fig. 5.10   Daily closing stock prices, with an anomaly that extends for
            several days (roughly from day 22) 80
Fig. 5.11   Distance of data point (12, 12) using one consecutive point
            on either side 83
Fig. 5.12   Distance of data point (12,12) from the regression line
            obtained using two consecutive neighbors on either side of it 85
Fig. 5.13   Synchronous rise and fall of $\mathscr{X}_1$ and $\mathscr{X}_2$ 88
Fig. 5.14   Asynchronous behavior of $\mathscr{X}_1$ and $\mathscr{X}_3$ 89

Fig. 6.1   Gaussian distribution with mean 0 and standard deviation =
            1; probability in the right tail (red area) is 0.0227 98
Fig. 6.2   Evaluation of $n(p, \alpha r)$ and $\hat{n}(p, r, \alpha)$ for a small dataset 103
Fig. 6.3   Clustering of six observations based on successive decrease
            in resolution values 105
Fig. 6.4   An example in which DB(pct, dmin)-outlier definition and
            distance-based method do not work well. There are two
            different clusters C1 and C2, and one isolated data object
            'a' in this data set. The distance from object 'b' to its
            nearest neighbor, d_2, is larger than the distance from object
            'a' to its nearest neighbor, d_1, preventing 'a' from being
            identified as an anomaly 107
Fig. 6.5   Anomaly detection in one dimension using a histogram.
            Bins centered at 3.0 and 4.2 consist of very few observations
            and therefore the associated observations may be considered
            anomalous 108
Fig. 6.6   Illustration of reachability distance. $d_{reach}(p1, o)$ and
            $d_{reach}(p2, o)$ for $k = 4$ 111
Fig. 6.7   An example in which LOF fails for outlier detection.............. 113
Fig. 6.8   Illustration of set-based path and trail definitions. The
            k nearest neighbors of p are $q1, q2$, and $q3$, for $k = 3$.
            Then SBN path from p is $\{p, q1, q3, q2\}$, and SBT is
            $< e1, e2, e3 >$ respectively 113
Fig. 6.9   Reverse Nearest Neighborhood and Influence Space.
            For $k=3$, $\mathcal{N}_k(q_5)$ is $\{q_1, q_2, q_3\}$. $\mathcal{N}_k(q_1) = \{p, q_2, q_4\}$.
            $\mathcal{N}_k(q_2) = \{p, q_1, q_3\}$. $\mathcal{N}_k(q_3) = \{q_1, q_2, q_5\}$.
            $\mathcal{RN}_k(q_5) = \{q_3, q_4\}$ 115

Fig. 7.1   Example of an outlier (Bob) identified by the fact that he is
            not very close to his best friend (Eric), unlike an inlier (Jack) 120

Fig. 7.2   Illustration of ranks: *Red dash* shows k-NN of p when
k is 3 and *long blue dash* shows a circle with radius of
$d(p, q3)$ and center of $q3$. The k-NN of p is {q1,q2, q3}.
Then $r_{q3}(p)=4$, because $d(q3, p)$ is greater than any of
$d(q3, q3), d(q3, q4), d(q3, q1)$ and $d(q3, q5)$ 121

Fig. 7.3   *Red dash circles* contain the k nearest neighborhoods of 'A'
and 'B' when $k=7$ 124

Fig. 7.4   Illustration of $d_k(p)$, $\mathcal{N}_k(p)$ and $\mathcal{RN}_k(p)$ for $k = 3$. The
large blue circle contains elements of $\mathcal{N}_3(p) = \{x, e, y\}$.
Because y is farthest third nearest neighbor of p,
$d_3(p) = d(p, y)$, the distance between p and y. The *smallest
green circle* contains elements of $\mathcal{N}_3(y)$ and the *red circle*
contains elements of $\mathcal{N}_3(z)$. Note that $r_y(p)$ = rank of p
among neighbors of $y = 9$. Finally, $\mathcal{RN}_k(y) = \emptyset$ since no
other object considers p in their neighborhood 125

Fig. 7.5   An example to illustrate 'Cluster Density Effect' on RBDA;
RBDA assigns larger outlierness measure to B 127

Fig. 7.6   Assignment of weights in different clusters and
modified-rank. Modified-rank of A, with respect to B, is
$1 + 5 \times \frac{1}{9} + \frac{1}{7}$ 129

Fig. 8.1   An example with three anomalous data points p, q, and r
added to synthetic data generated randomly 144

Fig. 8.2   Normalized LOF scores for data in Fig. 8.1; the y-axis
represents the data points 145

Fig. 8.3   Difference between the histograms of scores, (**a**) without
anomalies; and (**b**) with three anomalies 146

Fig. 8.4   AUC performance comparison of weighted adaptive
sampling (boxplots) with base algorithm LOF (*solid curve*),
for different values of k. (**a**) Wisconsin dataset. (**b**) Activity
dataset. (**c**) NBA dataset 149

Fig. 8.5   AUC performance comparison of weighted adaptive
sampling (boxplots) with base algorithm LOF (*solid curve*),
for different values of k (Continued). (**a**) KDD dataset. (**b**)
PEC dataset 150

Fig. 8.6   AUC performance vs. number of iterations. (**a**) Synthetic
dataset. (**b**) Real dataset 151

Fig. 9.1   **Time series representing normalized daily closing stock
prices for some companies**–Time series representing
normalized 154

Fig. 9.2   Plots of trend $(\text{sgn}(\Delta x(t) = x_{t+1} - x_t))$ for two time series.
These two series move in the same direction five out of eight
times 161

Fig. 9.3   **Two time series compared using DIFFSTD**—*Vertical
            lines* indicate differences at individual time points 162
Fig. 9.4   Time alignment of two time series; aligned points are
            indicated by the *arrows* 163
Fig. 9.5   **Example for SAXBAG distance computation**—SAX
            words and SAXBAG frequency counts 165
Fig. 9.6   Examples for abnormal subsequences: Subsequences
            extracted from a long series of power consumption for a
            year; each subsequence represents power consumption for
            a week. (*Top left*) contains normal subsequences; others
            contains abnormal subsequences along with the week they
            appear. Abnormal subsequences typically represent the
            power consumption during a week with special events
            or holidays 177
Fig. 9.7   *Left* (No reordering): 6 calculations performed before
            abandoning; *Right* (With reordering): Abandoning after 3
            largest distance computations 179
Fig. 9.8   **Sub-sequence frequency**—*Red bold line* represents a
            subsequence considered abnormal since it appears only
            once in this series, whereas other subsequence patterns
            occur much more frequently 179
Fig. 9.9   Performance of SAXFR versus length of the subsequence.
            In general, the ratio of anomaly scores for abnormal
            subsequence to normal subsequence decreases with the size
            of the subsequence 180
Fig. 9.10   NMUDIM anomaly scores for each data set. Plots of the
            online anomaly scores for each series at time $100+x$ in
            four data sets (*from top to bottom, left to right*) : SYN2,
            STOCKS, MOTOR, and SHAPE1. *Red curves* indicate
            anomalous series 185
Fig. 9.11   Experimental results for VIDEOS1. *Red circle* highlights
            abnormal subsequences. (*Top Left*) Plot of VIDEOS1
            time series; (Other) Results of 6 algorithms used in these
            comparisons. Y-axis represents anomaly scores at time t.
            X-axis shows time t 187

Fig. A.1   A synthetic dataset with clusters obtained by placing all
            points uniformly with varying degrees of densities 192
Fig. A.2   A synthetic data set with one cluster obtained using the
            Gaussian distribution and other clusters by placing points
            uniformly 192

Fig. B.1   **Typical time series problems**—Two series are $x_a(t)$
            and $x_b(t)$. Time lagging: $x_a(t) = x_b(t + x)$; Amplitude
            differences: $x_a(t) <> x_b(t)$ 200

Fig. B.2   Plots of first six time series dataset of Table B.2. *Red dotted
             line* represents anomalous series 200
Fig. B.3   Plots of time series 7 to 12 of the dataset of Table B.2. *Red
             dotted line* represents anomalous series 201
Fig. B.4   Four datasets used in the experiments (*from top to bottom,
             left to right*) are: SYN2, STOCKS, MOTOR, and SHAPE1.
             Anomalous series are marked as *red* 202
Fig. B.5   Experimental results for SYN0. *Red circles* highlight
             abnormal subsequences. (*Top Left*) Plot of SYN0 time
             series; (Other) Results of 6 algorithms used in these
             comparisons. Y-axis represents anomaly scores at time *t*.
             X-axis shows time *t* 204
Fig. B.6   Experimental results for ECG1. *Red circles* highlight
             abnormal subsequences. (*Top Left*) Plot of ECG1 time
             series; (Other) Results of 6 algorithms used in these
             comparisons. Y-axis represents anomaly scores at time *t*.
             X-axis shows time *t* 204
Fig. B.7   Experimental results for ECG2. *Red circles* highlight
             abnormal subsequences. (*Top Left*) Plot of ECG2 time
             series; (Other) Results of 6 algorithms used in these
             comparisons. Y-axis represents anomaly scores at time *t*.
             X-axis shows time *t* 205
Fig. B.8   Experimental results for ECG2. *Red circles* highlight
             abnormal subsequences. (*Top Left*) Plot of ECG2 time
             series; (Other) Results of 6 algorithms used in these
             comparisons. Y-axis represents anomaly scores at time *t*.
             X-axis shows time *t* 205
Fig. B.9   Experimental results for TK160. *Red circles* highlight
             abnormal subsequences. (*Top Left*) Plot of TK160 time
             series; (Other) Results of 6 algorithms used in these
             comparisons. Y-axis represents anomaly scores at time *t*.
             X-axis shows time *t* 206
Fig. B.10   Experimental results for TK170. *Red circles* highlight
             abnormal subsequences. (*Top Left*) Plot of TK170 time
             series; (Other) Results of 6 algorithms used in these
             comparisons. Y-axis represents anomaly scores at time *t*.
             X-axis shows time *t* 206
Fig. B.11   Experimental results for VIDEOS2. *Red circle* highlights
             abnormal subsequences. (*Top Left*) Plot of VIDEOS2
             time series; (Other) Results of 6 algorithms used in these
             comparisons. Y-axis represents anomaly scores at time *t*.
             X-axis shows time *t* 207

List of Tables

Table 6.1   Illustration of ROF evaluation for observations in Fig. 6.3 105

Table 7.1   Performance of each algorithm for synthetic dataset 2. The
            largest values (best performances) are shown in red color 131

Table 7.2   Summary of LOF, COF, INFLO, DBCOD, RBDA,
            RADA, ODMR, ODMRS, ODMRW, and ODMRD for all
            experiments. Numbers represent the average performance
            rank of the algorithms; a larger value implies better
            performance. Data set with 'r' in the parentheses represents
            the data set with rare class. And data set with 'o' in the
            parentheses represents the data set with planted outliers 133

Table 7.3   KDD 99, $k = 55$. For each value of m the best results are
            in red color. Results for COF and INFLO were generally
            poorer than for LOF and are not shown below 133

Table 7.4   Packed Executable Dataset $k = 55$. For each value of m the
            best results are in red color. Results for COF and INFLO
            were generally poorer than for LOF and are not shown below 134

Table 8.1   Performance comparison: averaged AUC over 20 repeats 150
Table 8.2   Performance over Sum of AUC rankings with different
            combination approaches 151

Table 9.1   Pros and Cons of using various distance measures for
            anomaly detection in time series 170
Table 9.2   RankPower comparison of algorithms; using 47 data sets 184
Table 9.3   Performance of all algorithms; numbers show the true
            outliers that are identified by algorithms 186
Table 9.4   Running time of NMUDIM and average computation
            workload comparison between NMUDIM and UMUD 188

Table B.1   Time series $\mathscr{X}$ used in SXBAG evauation   196
Table B.2   Summary of all time series sets. Similarity of normal series
            represents that how similar normal series look like to each
            other. N represents number of series and O number of
            outliers   197
Table B.3   Summary of all time series sets (Cont)   199
Table B.4   Time series data sets details   203

Part I
Principles

Chapter 1
Introduction

Incidents of fraud have increased at a rapid pace in recent years, perhaps because very simple technology (such as email) is sufficient to help miscreants commit fraud. Losses may not be directly financial, e.g., an email purportedly from a family member may pretend to communicate a photograph, clicking on whose icon really results in malware coming to reside on your machine. As is the case with health and other unpreventable problems faced by humanity, early detection is essential to facilitate recovery. The automated detection and alerting of abnormal data and behaviors, implemented using computationally efficient software, are critical in this context. These considerations motivate the development and application of the anomaly detection principles and algorithms discussed in this book.

A central aspect of several instances of crime is that scammers take advantage of a potential of humans to confuse the *plausible* and *possible* with the *not improbable*. It is certainly possible that someone's relative is in jail, but if it hasn't happened before, and is not consistent with past behavior or experiences of that relative, then one should consider the jailing of the relative (and his sudden need for money) to be an *abnormal* or *anomalous* event that cannot be taken at face value. Based on past experiences of multiple individuals, the probability of being scammed is much higher than the probability of one's relative suddenly being jailed in a foreign country and needing money. Thus, the ability to identify anomalies, especially in relation to the past history of oneself and others, is critical to one's financial and other kinds of safety.

Substantial efforts are expended towards detecting when attacks and other problems occur, based on the principle that abnormalities in underlying behavior manifest themselves in observations that can be used to separate them from "normal" (non-attack) modes of behavior. The broad approaches in detection include:

- Look for damage after damage has occurred (e.g., look for unexpected transactions in a credit card bill);

© Springer International Publishing AG 2017
K.G. Mehrotra et al., *Anomaly Detection Principles and Algorithms*, Terrorism, Security, and Computation, https://doi.org/10.1007/978-3-319-67526-8_1

- Detect signs of damage (e.g., less data space available on hard drive than expected);
- Pattern-match against known "signatures" of malware, Internet site addresses, message routing paths, etc.; and
- Anomaly detection: compare against expected or normal behaviors or data.

Detection of anomalies is the focus of this book, and the next section discusses the general principles associated with anomaly detection. Following detection, actions can be taken to recover from the attack. These may include patching software holes, deleting undesirable files, changing passwords, changing hardware, etc.; however, such recovery aspects are application-dependent, and hence out of the scope of the contents of this book.

The rest of this chapter discusses what we mean by an anomaly, and describes some of the application areas wherein anomaly detection is essential. Later chapters address anomaly detection algorithms in greater detail.

1.1   What's an Anomaly?

Anomalies or *outliers* are substantial variations from the norm.[1,2]

Example 1.1   The results of an IQ test are expected to be around 100, with a standard deviation (s.d.) of 15, as shown in Fig. 1.1. If an individual's IQ test result is 115, it varies from the norm, but not substantially, being only about one s.d. higher. On the other hand, an individual whose IQ test result yields a score of 145 would be considered anomalous since this value is higher than the mean by about three times the s.d. This example is relatively straightforward since there is a single quantitative attribute (IQ score) with a unimodal distribution (with well-known statistics) that can be used to identify anomalies. Most real problems, addressed by anomaly detection algorithms, are multi-dimensional, and may involve nominal or categorical (non-numeric) values.

Anomaly detection approaches are based on models and predictions from past data. The primary assumption of normal behavior is stationarity, i.e., the underlying processes, that led to the generation of the data, are believed not to have changed significantly. Hence the statistics that characterized a system in the past continue to characterize the system in the future; in other words, what we have seen before is what we expect to see in the future. Data that changes over time, in some cases, may be characterized by long-term trends (e.g., increasing heights or lifespan among humans), or by cyclic behavior.

---

[1]Some authors refer to "anomalies" in processes and "outliers" in data. This book uses the two words synonymously.

[2]Aspects of anomaly detection are also studied under other names, such as 'novelty detection', 'chance discovery', 'exception mining', and 'mining rare cases'.

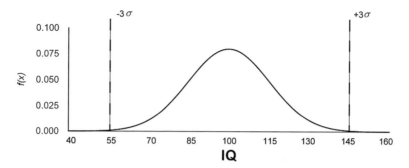

Fig. 1.1 Distribution of Intelligence Quotient (IQ) scores—Mean of 100, Standard Deviation of 15

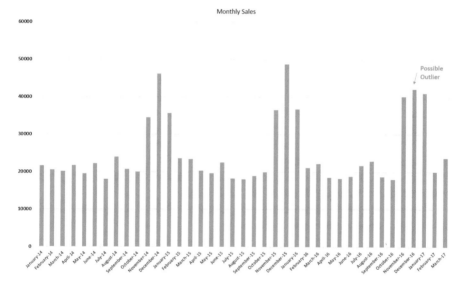

Fig. 1.2 Monthly sales for a retailer—December 2016 sales are anomalous with respect to prior year December sales

Example 1.2 Consider the monthly sales of a retailer as shown in Fig. 1.2. Sales in December 2016 may substantially differ from those in November 2016 and January 2017, due to the effect of a large volume of goods purchased during the Christmas season. Those values may seem anomalous with respect to each other, but are not anomalous with respect to the same months in preceding years. In order to make predictions about January 2017, we need to rely on numbers from January months in 2014–2016, while factoring in year-to-year trends (e.g., steady growth or decline). In some such examples, we can rely on common-sense knowledge or a *model* of the underlying process, which includes seasonality as well as annual trends. In other cases, the first task is to identify such a model or process, or at least to characterize the effects of an unknown process on observable data, so that one can estimate the likelihood with which a particular data point could have been generated by the underlying (unknown) process, based on a characterization of what is reasonable (or non-anomalous) from other observable data generated by the same process. With monthly retail sales data, for instance, well-known time series modeling techniques can uncover the trends and cycles underlying the data, even if we are not told in advance that they represent retail sales; hence, even if December 2016 sales had been approximately the same as November 2016 sales, they may be flagged as anomalous (and alarming) if they are much lower than December 2015 sales. The degree of *anomalousness* of a data point is thus dependent on the entire data set available to us, not just the very recent data.

At first sight, it appears that the problem is one of classification, i.e., separating data into two classes: anomalous and non-anomalous. It is tempting to address this problem using well-known machine learning and classification algorithms, e.g., back-propagation neural networks, support vector machines, and decision trees (*a la* CART). This, however, will rarely be successful since there is a drastic imbalance between the two classes: anomalous data are much rarer than non-anomalous ones; results obtained by classification algorithms will often result in too many false negatives (i.e., not recognizing anomalies). Further, the various anomalous cases may have very little in common. Finally, the occurrence of an anomaly may well be within the same bounds as those characterizing non-anomalous data, and hence not distinguishable directly by attribute values, but may require analysis of their behavior with respect to subsets of neighbors or other data points (e.g., December 2016 using December 2015 and December 2014). This motivates the development and application of carefully designed anomaly detection algorithms, along with an understanding of their applicability and limitations.

The next chapter addresses critical questions relevant to the formulation of anomaly detection algorithms:

- How is the norm characterized?
- What if there are multiple and substantially different cases, all of which can be considered to be normal?
- What is considered to be a substantial variation, as opposed to a minor variation from a norm?
- How do we address multi-attribute data?
- How do we address changes that happen over time?

In the rest of this chapter, we describe examples of application areas where anomaly detection is useful, and sometimes necessary to perform certain critical tasks. The applications mentioned here are cybersecurity, finance, health, defense, home safety, industry, and science; several other application areas also utilize anomaly detection algorithms.

1.2 Cybersecurity

There are many aspects of cybersecurity, and the rest of this section discusses those in which anomaly detection plays a critical role. We first present a motivating example that illustrates the complexity of the problem.

Consider the example in which a criminal hacks into a normal user's account first; he then observes the user's behavior, and eventually pretends to be the same user in the system before launching malicious attacks. After surfing the system and identifying weaknesses, a series of attacks are triggered to hack the system, to steal data or achieve other objectives. During this entire process, each single attack may look normal, but when the individual commands are linked together, an abnormal pattern may be detectable. For example, if the number of times a system file is explored is too high, compared with other users (especially the folder belonging to administrators), that may indicate anomalous behavior. Similarly, anomalous behavior may be characterized by the occurrence of an unusually high number of system logins and logouts during non-working hours, or increased use of remote "ssh" commands to connect to the system; but such behaviors can be inferred only if observed over a relatively long period of time.

1.2.1 Privacy

Details of the bank accounts and medical records of Hollywood celebrities have recently been exposed, with the ensuing loss of privacy of those individuals. Selfies belonging to many Hollywood celebrities have been shared publicly on the Internet because owners' iCloud accounts have been hacked. These are instances of violations of data privacy.

Confidential data private to individuals (such as family details, and medical conditions) maybe non-financial but can eventually be used to gain unauthorized access to financial assets. For example, information such as one's mother's maiden name and social security number are frequently used to permit access to individuals who may not have the passwords required to electronically access a bank account. Other confidential information may be potentially embarrassing if revealed, so that those who possess such information may extort a fee for remaining silent about the same.

Access control mechanisms are usually in place but their implementation is often flawed, with a false sense of security that is readily breached. Although it is practically impossible to prevent inappropriate access (since individuals with access may commit crimes or unknowingly provide access to others), measures can be taken to identify rapidly the possible occurrences of privacy violations. Anomaly detection algorithms can be used to monitor access to the data, and flag variations from the norm, e.g., identifying when individuals (who may have access permissions) access data at time points that are anomalous when compared to their prior access histories and the access histories of others with similar roles in the same organization.

Loss of individuals' private data may also lead to attacks on organizations where the individuals are employed, with the potential theft of trade secrets and intellectual property. This may eventually lead to the loss of markets and drastic reduction in profitability, if competitors produce the same products at a lower price, since they do not have to pay for the research costs undergone by the original organization. Detecting anomalies in access patterns, along with the frequencies and patterns of email and other communications of employees may assist in discovering potential loss of confidential data.

1.2.2 Malware Detection

The traditional approach to detecting viruses and worms focuses on after-the-fact recognition of their signature patterns, and looking for such patterns within program code and data. However, this approach is inadequate in dealing with the sudden influx of a new malware instance that does not match old malware signature patterns. An anomaly detection approach would instead monitor the appearance and behavior of malware to attempt to recognize variations from the norm.

1.2.3 Fraudulent Email

The Internet is periodically flooded with email messages that claim to originate from financial, information technology, or other service providers with whom individuals routinely conduct business. Many individuals are lulled into clicking on apparently harmless websites, or providing their personal data to others who then participate in fraudulent activities, or sell the personal data on the *darknet*. Anomaly detection algorithms can be used to guard against such attacks, at the individual level as well as by organizations protecting their users.

At the organizational level, the frequencies of messages from different sources (IP addresses) can be monitored, and the variations in these may be flagged as potential problems; similarly, anomalies may be detected in message sizes, based on prior size distribution information. At the individual level, every email message that doesn't conform to some generic expectations can be flagged as potentially harmful, e.g.,

- messages from unknown senders,
- messages that arrive at an unusual time of day,
- messages containing unusual typographical errors,

- messages in which the form of salutation differs from previous forms of salutations used by the same sender in the past,
- messages which ask the user to click on a link, and
- messages which explicitly ask the user to forward the message to others.

Each of the above is an anomaly detection task, focused on a different aspect of the message.

1.3   Finance

Many cybersecurity attacks are focused on obtaining or misusing financial data and resources of gullible individuals. However, such crimes are also committed using other means, e.g., hardware or software installed at automated teller machines (ATMs). The following are a few examples of the same.

1.3.1   Credit Card Fraud

Unauthorized usage of credit cards and debit cards is a major problem leading to the theft or loss of billions of dollars every year. For example, in 2013, newspapers reported that a very large retail and financial firm had been the subject of a major incident of theft of millions of credit card numbers. Exact amounts of losses are unknown, but the security of 40 million cards was compromised[3]; an additional 70 million people also suffered from potential breach of private data (such as mailing address). Other well-known incidents include the 2007 card compromise that affected 90 million customers, and the 2009 compromise affecting 130 million credit cards. A 2016 research study predicted that the cost of data breaches will increase to $2.1 trillion by 2019, almost four times more than the global cost of breaches in 2015 [89].

Banks and credit card companies take many measures to facilitate the detection of potential anomalies in credit card usage, e.g., the use of a card in a location geographically distant from the specific user's normal usage region, or to make purchases of unusual items such as electronic equipment using amounts of money that are unusual for that user. Yet many early fraudulent usage instances go undetected, especially if they involve small amounts; the danger here is that these small amounts may indicate a test, followed soon by a large fraudulent purchase using the same card. Regular and periodic application of anomaly detection algorithms on recent purchase data would help prevent such problems to some extent.

---

[3]The word "compromise" has come to acquire a technical meaning: the simultaneous loss of private information for a large number of individuals, in a single data theft incident.

1.3.2 Creditworthiness

An important source of revenue for banks is the interest on the loans they make to individual consumers or businesses. Many of these loans are of relatively low risk, and banks expect that the loans and interest will be repaid without problems. However, a small number of these loans are not repaid, and lead to significant losses for the banks, which are eventually passed on to other consumers, e.g., in increased interest rates. Even when collateralized (e.g., home mortgages), banks incur significant costs in the processes needed to claim the mortgaged property and then monetize the assets. Hence it would be useful if the risk of defaulting on a loan is substantially reduced at the time of making a loan; at the same time, qualified people should not be denied loans (as has occasionally happened in the past with racially biased decision-making). Accurate anomaly detection on the credit history and other data from loan applicants is hence desirable.

1.3.3 Bankruptcy Prediction

Risk is inherent in any entrepreneurial venture, and a significant number of companies file for bankruptcy every year, affecting the company's owners (or stockholders), employees, contractors and creditors. Detecting potential bankruptcy at an early stage would be very helpful to all these stakeholders. Anomaly detection algorithms have been applied successfully to the task of analyzing company fundamentals (such as earnings) over time, to evaluate which companies are likely to go bankrupt.

1.3.4 Investing

Stock prices fluctuate every day, and the investors' holy grail is to predict the future performance of a stock, making buy/sell/hold decisions on that basis. Accurate prediction is practically impossible, with unpredictable market shocks that affect stock prices. However, the performance of a stock over a short (recent) period of time can be compared with its prior performance, as well as the performance of other stocks of similar companies. This can help identify anomalies that may signify that the company is outperforming or underperforming its competitors. Thus the application of anomaly detection algorithms can provide valuable information to potential and current investors in the company.

1.4   Healthcare

Many aspects of human healthcare can be assisted by anomaly detection algorithms; two primary areas are the diagnosis and monitoring of patients.

1.4.1   Diagnosis

Practically every diagnostic effort is based on data that shows abnormalities in the patient's behavior or *vitals* (easily observable data characterizing the patient, e.g., blood pressure). In some cases, accurate analysis of quantitative data describing the patient is non-trivial, and would benefit from the application of anomaly detection algorithms. A few examples are described below.

- EKG and EEG data: Electrocardiogram (ECG/EKG) data are time series that measure the electrical activities of the heart. Arrhythmias are irregularities in electro-cardiogram data, e.g., premature ventricular contraction (PVC) is a relatively common cardiac arrhythmia that can be detected in EKG data by comparison with normal EKG data. Automatic detection of PVC from EKG data would significantly reduce the workload for cardiologists, and potentially increase the detection rate of PVC. Arrhythmias are currently detected by visual inspection conducted by a physician expert who may not always be at hand when a patient is ill with cardiac problems; this task could potentially be performed more efficiently by anomaly detection algorithms. Automation can also help reduce potential human errors by overworked physicians.

  A similar issue is that of identifying possible evidence of epileptic seizures in patients, by examining electro-encephalogram (EEG) data for abnormal variations; the scarcity of qualified neurologists suggests that the use of anomaly detection algorithms could be helpful. In such cases, the anomaly detection algorithms could at least serve triage purposes: a certain number of false positives can be tolerated by a physician as long as the number of false negatives is near-zero.

- Cancer Diagnosis: The classification of tumors as benign *vs.* malignant, from radiographic image data, has long been known to be a task that is particularly challenging because of the relatively small number of malignant cases. Machine learning and classification algorithms have been applied to this problem. Anomaly detection algorithms could be very helpful in this application, since various malignant cases may not have much in common, beyond the fact that their attributes differ significantly from benign cases. The hardest cases, of course, are the malignant cases whose attribute values are indistinguishable from the benign cases; identifying potential features that may help to distinguish them would be helpful, prior to application of classification or anomaly detection algorithms.

1.4.2 Patient Monitoring

For patients being treated for some serious health disorders, it is very important to monitor progress and vital signs constantly, in order to determine the occurrence of unexpected side effects of medications or surgery, requiring immediate attention and additional emergency treatment in some cases. Unfortunately, due to the large number of patients in a hospital, signs of an abnormality are sometimes accidentally ignored or do not receive immediate attention. Similarly, elderly and disabled individuals occasionally suffer from falls in their private residences, without receiving immediate care. The application of anomaly detection algorithms to alert care providers, based on appropriate sensors, is hence essential.

1.4.3 Radiology

The field of radiology often involves searching for unusual data in X-ray, NMR, and other images. Anomaly detection algorithms can potentially assist in finding early phase tumor, facilitating early detection of cancer.

1.4.4 Epidemiology

Viruses and bacteria evolve at a fast pace, and understanding them is vital for at least temporary success in the arms race between medications and pathogens. The fields of genetics, proteomics, and metabolomics can be assisted by anomaly detection algorithms that may search for unusual mutations that may signal specific diseases. Medical science would also benefit by identifying points in time at which epidemiological data reveals that previously successful medications cease to be helpful to patients, signifying the emergence of a drug-resistant mutation of the responsible pathogen. Some such data can also be obtained from individual patients, whose response to a medication may follow an unusual path, e.g., first appearing to improve, then degrading rapidly. This may call for actions such as quarantine procedures to prevent the epidemic spread of a new drug-resistant pathogen.

1.5 Defense and Internal Security

Defense and internal security application areas are similar to the problems faced in the field of cyber-security, with the main difference being that observations may be made of the behaviors of individual people or physical sensors.

1.5.1 Personnel Behaviors

Unusual behaviors of people in public places (such as airports) may indicate intent towards planned violent activities. The behaviors may be observed using video-cameras with limited regions of surveillance, installed in multiple public areas, and monitored by security personnel. Based on the frequent observed behaviors of people, normal patterns may be established with respect to how people tend to move, and variations from these patterns represent anomalies which can be detected and investigated. The application of automated anomaly detection algorithms to such video data is a non-trivial task, especially since human subjects may move between the regions monitored by different cameras, making it difficult to establish a clear narrative regarding each individual subject's behavior over an extended period of time. There is also a potential for many false positive reports, since anomalous behavior may be symptomatic of illness or confusion on the part of the subjects, rather than violent intent. Indeed, the latter cases are extremely rare, making their characterization difficult. An additional difficulty is that individuals with violent intent may be well aware that they are being monitored, and hence take extra care to appear 'normal' until an actual violent event begins to be carried out.

1.5.2 Battlefield Behaviors

In the military scenario, efforts are made to infer the tactical intent of an opposing party from observable behaviors. Several classical battles have been won or lost when one party conducts unusual movements of forces which are not foreseen by the other party, even when the latter has greater strength in numbers or equipment. For instance, an ambush or feint or a pretend attack by the first party with a small number of forces may be intended to make the second party respond in a predictable way which is advantageous to the first party which plans on a tactic that has not been considered by the second party. Models would be constructed by each party regarding possible intentions of the opposing party, evaluating the most likely courses of action that may be taken. Variations from these expectations should lead to questioning the model, and evaluation of alternative models to see if they would fit better with observed data. If none of the current models fit well with the data, closer monitoring and readiness to take actions would be required.

1.5.3 Unconventional Attacks

Although there is broad agreement around the world on avoiding chemical and nuclear warfare, there have been some occasions where these expectations have been violated by individual governments or factions. The potential for their use hence cannot be ruled out, requiring monitoring of signs indicating the accumulation of such agents as well as their use or experimentation, e.g., byproducts of chemical reactions or radioactive decomposition. Some such data may come from chemical monitoring stations, while others may come from satellite images. In some cases, the relevant anomalies may be detected based on evidence of unusual activities, such as the movements of individual scientists or high level military officers to unexpected locations. As in the case of malware, looking for a specific pattern may not yield sufficient information, but looking for variations from known *normal* behaviors may be much more helpful.

1.6   Consumer Home Safety

Anomaly detection algorithms, assisted by sensors that are becoming ubiquitous in the home, can help with problems encountered in many households, some of which are discussed below.

1.6.1   Detecting Occurrence of Falls and Other Problems

Many disabled and senior citizens live alone or in environments with very little day-to-day human contact and direct monitoring. Disabilities as well as weakened bone structures frequently result in falls and medical emergencies from which they may be unable to recover by themselves. Unfortunately, those who can help them may not have scheduled visits to occur close to the time at which such incidents occur, so that easily addressable problems escalate in intensity to life-threatening situations. Some simple technologies, e.g., using accelerometers embedded in smart phones, can help detect some such situations, but do not address other scenarios where falls tend to occur, e.g., in bath-rooms or bed-rooms where individuals do not carry smart phones on their bodies. However, strategically placed vibration and acoustic sensors may help in such situations, accompanied by anomaly detection algorithms that trigger alarms (with medical personnel or relatives) only when the sensed data indicates abnormal behavior, varying significantly from characteristics of data collected over a prior period of time during which no falls are known to occur. The efficacy of this approach depends critically on the accuracy of the anomaly detection algorithms: too many false positives would be annoying since they result in excessive caregiver resource consumption and desensitization to true alarms; on the other hand, even a single false negative (undetected fall) would degrade confidence in the usefulness of the system, negating its utility.

1.6.2   Home Perimeter Safety

Signs of break-ins and burglaries are now routinely detected by sensors placed near windows and doors, and can be monitored by security organizations, at costs that vary from provider to provider. However, false alarms sometimes occur, resulting in wasted resources, as in the case of false positives in fall detection (discussed above). It may then be reasonable to use evidence from multiple sensors to confirm possible break-ins, before deploying human personnel and law enforcement officers to the possible scene of a crime. For instance, we may expect that if a sensor in one location is tripped by an intruder, this would be followed closely by another, assuming humans are moving at a reasonable pace through the house. On the other hand, if a sensor was tripped by a tree branch, a squirrel or a deer on the lawn close to a window, we would expect that the sensor behavior pattern would be quite different, e.g., exactly one sensor may be tripped, possibly multiple times in quick succession. This calls for the application of pattern recognition and anomaly detection algorithms, with data collected from "normal" and "abnormal" behavior that may be simulated experimentally.

Additionally, in regions where the incidence of crime is relatively high, periodic monitoring of local traffic and pedestrian behaviors may reveal "normal" patterns of behavior within a specific neighborhood, variations from which may trigger additional monitoring or deployment of law enforcement resources. For instance, people living in a neighborhood may generate an entry into a neighborhood database whenever they see an unknown vehicle enters their neighborhood, and these may be linked to a law enforcement database that keeps track of stolen vehicles or those suspected of involvement in other crimes. Multiple sightings of the same vehicle or individuals, assisted by an anomaly detection algorithm, may trigger an investigation into the possibility of a break-in being planned.

1.6.3   Indoor Pollution Monitoring

Many sensors currently exist to monitor the concentrations of carbon-monoxide and volatile organic compounds. In large cities, it is also important to monitor the levels of particulate air pollutants and sulfur-di-oxide. The effective use of such sensors would be in collusion with anomaly detection algorithms that can sense potential problems before they actually occur, e.g., using information from external sensors, weather data (e.g., wind velocity), and variations in pollutant density gradients, along with data relevant to "normal" conditions.

1.7   Manufacturing and Industry

Anomaly detection algorithms can be particularly useful in manufacturing and industry scenarios since they can help catch potentially serious problems at a very early stage.

1.7.1   Quality Control

Statistical *change detection* algorithms have been used for a long time for quality control in manufacturing organizations, triggering alarms when sampled output characteristics of a product fall below expected quality constraints. In general, fluctuations in the underlying process may be detected by drastic changes in specific sensor data measurements. These may be considered to be simple anomaly detection algorithms.

In addition, anomaly detection can be applied to data from multiple sensors located at various points in the monitored manufacturing environment. In addition to identifying problems after they have occurred, anomalies may be detected in unusual patterns of various sensor data, indicating possible locations in the manufacturing environment where faults or failures have occurred. For example, if the "normal" behavior of two adjacent sensors (possibly measuring different attributes or features of the system or its products) over time involves a linear relationship, with previously measured gradients, then a significant variation in this relationship may be detected as anomalous, triggering further investigation.

1.7.2   Retail Sales

Many retail organizations have to constantly monitor their revenues and earnings, to facilitate planning as well as to identify any potential disasters at an early stage. This involves constructing time series of sales data, and analyzing fluctuations in sales, comparing them to prior values, while factoring in various trends and relationships to periodic cycles and external events. Anomaly detection algorithms can play a useful role in this context, helping to separate insignificant fluctuations (noise) from potentially meaningful variations with significant implications for the future plans of the organization.

1.7.3   Inventory Management

Many retail and other organizations maintain inventories of various products and raw materials, and their effective management is an important factor influencing profitability, especially when demand fluctuates drastically over time. Difficulties arise when an organization does not have a product available for sale when there is a sudden surge in demand; conversely, maintaining extra inventory (in anticipation of a demand surge) can be expensive. Finally, some organizations are plagued by occasional occurrence of theft by employees or outsiders, which may only be detectable by effective inventory management and monitoring. Anomaly detection algorithms can play a role in this context, e.g., by enabling formulation of mathematical models or case data that describe "normal" behavior of inventory data (collected over time), and triggering warnings when such expectations are significantly violated.

1.7.4   Customer Behavior

Most organizations are constantly striving to determine how best to allocate resources to maximize profit or revenue, by understanding customer behaviors. Anomaly detection algorithms can be considered to be one aspect of data mining efforts in this context. Past data can be analyzed to model customers' typical purchasing behaviors, and analyzing the subclasses of behaviors wherein purchasing increases or decreases with time, perhaps as a result of a change in store configuration or website. The application of anomaly detection algorithms can enable detecting variations from such models, triggering investigations into probable causes and possible remedies.

In addition to purchasing, customer behaviors may also be relevant to identify potentially unacceptable actions or deception, e.g., in applications such as money laundering. Anomaly detection algorithms can then pursue three directions: comparing an individual's behavior to his own past behavior, or to the behavior of others in the group or category to which he belongs, or to the behavior of the entire collection of customers. Such monitoring may also reveal fraud being perpetrated by other individuals, who may have gained unauthorized access to a customer's account.

1.7.5   Employee Behavior

Many organizations performing sensitive actions need to be extremely sensitive to the possible damage caused by a few errant employees who may have access to organizational resources in the course of normal performance of their everyday jobs. Indeed, some of the largest known fraudulent financial manipulations have been identified as occurring due to the unusual activities of a small number of individual employees, which could be detected by the passive monitoring of their actions using anomaly detection algorithms.

External agencies can also apply anomaly detection algorithms to determine whether fraud is being perpetrated by the principals of an organization. For example, a famous Ponzi scheme could have been detected early by investigators if they had analyzed the promises of high guaranteed returns made by the organization to investors; such promises are well outside the norms of other stockbrokers and hedge funds, including even the most successful members of this group. The promise of high returns along with substantial opacity in the investing philosophy of the organization should have triggered warning bells.

In addition to financial organizations, retail stores must monitor their employees to maintain productivity and integrity; this may again be assisted by anomaly detection algorithms.

1.8   Science

The application of anomaly detection algorithms is ubiquitous in science. Indeed, according to one perspective, progress in science occurs due to paradigmatic revolutions caused by the discovery of anomalous data that contradict well-established models [78]. We consider below a few examples of anomaly detection in everyday scientific practice, rather than revolutions.

The SETI project involves large scale efforts utilizing thousands of computers that have been launched to analyze electromagnetic data received by the earth, searching for anomalies that may indicate possible transmission of meaningful signals by intelligent extra-terrestrials. More successful have been efforts applied in the search for planets and stars with unusual behavior (compared to most other objects), revealing the existence of planets whose temperature and composition enables the occurrence of liquid water, hence presumed to be hospitable to life similar to that on earth.

More routinely, the not-so-remote skies are periodically scanned with telescopes to discover any unusual objects that do not fall into the categories of known objects such as satellites; such monitoring is conducted to evaluate potential threats from other nations as well as natural objects in space that may be heading towards the earth. Even when they do not approach the earth, large natural objects flying through space have allowed us to learn much about the universe and its composition.

Other scientific applications of anomaly detection include the search for new subatomic particles, the discovery of new species from paleontological records, and the detection of new strains of disease-causing organisms.

1.9 Conclusion

This chapter presented the motivation and several application areas for anomaly detection algorithms. There is a great need to develop general-purpose as well as problem-tailored anomaly detection techniques, which must be adapted to keep pace with the latest changes in technology which may result in new vulnerabilities in various systems. Anomaly detection techniques must be efficient enough to catch the small amount of outliers within large data flows, and must also be smart enough to catch anomalies over time periods that may be short or long.

The need for anomaly detection algorithms goes well beyond direct financial losses to individuals or companies. An inability to detect cyber-attacks and financial fraud can lead to the suspicion that our existing financial and other infrastructures are not safe, since it is unknown whether serious financial crimes are being conducted right under our noses. This leads to the fear that the whole economic system may come crashing down at any minute, thanks to the combination of rogue actors, impotent institutions, careless system-builders, and ignorant individuals.

In the next few chapters, we discuss general principles underlying anomaly detection. This is followed, in Part II, with the details of specific anomaly detection algorithms, some of which have been developed in the very recent past.

Chapter 2
Anomaly Detection

Anomaly detection problems arise in multiple applications, as discussed in the preceding chapter. This chapter discusses the basic ideas of anomaly detection, and sets up a framework within which various algorithms can be analyzed and compared.

2.1 Anomalies

An anomaly is a "variation from the norm"—this section explores this notion in greater detail.

Many scientific and engineering fields are based on the assumption that *processes* or *behaviors* exist in nature that follow certain rules or broad principles, resulting in the *state* of a system, manifested in observable *data*. From the data, we must formulate hypotheses about the nature of the underlying process, which can be verified upon observation of additional data.[1] These hypotheses describe the *normal* behavior of a system, implicitly assuming that the data used to generate the hypotheses are typical of the system in some sense.

However, variations from the norm may occur in the processes, hence systems may also exist in *abnormal* states, leading to observable data values that are different from the values observed when no such process/state variations occur. The task of anomaly detection is to discover such variations (from the norm) in the observed data values, and hence infer the variations in the underlying process. A fundamental problem is that there is no simple unique definition that permits us to evaluate how similar are two data points, and hence how different is one data point from others in the data set. As pointed out by Papadimitriou et al. [92],

---

[1] In some cases, the inputs to a process can be perturbed, and the resulting changes in the output data observed.

© Springer International Publishing AG 2017
K.G. Mehrotra et al., *Anomaly Detection Principles and Algorithms*, Terrorism, Security, and Computation, https://doi.org/10.1007/978-3-319-67526-8_2

"...there is an inherent fuzziness in the concept of *outlier* and any outlier score is an informative indicator than a precise measure."

When an anomaly detection algorithm is applied, three possible cases need to be considered:

1. *Correct Detection:* Detected abnormalities in data do correspond exactly to abnormalities in the process.
2. *False Positives:* The process continues to be normal, but unexpected data values are observed, e.g., due to intrinsic system noise.
3. *False Negatives:* The process becomes abnormal, but the consequences are not registered in the abnormal data, e.g., due to the signal of the abnormality being insufficiently strong compared to the noise in the system.

Most real-life systems are such that 100% correct detection is impossible. The task of the data analyst is to recognize this fact, and devise mechanisms to minimize both false positives and false negatives, possibly permitting more of one than the other due to considerations of asymmetrical costs associated with these.[2]

To account for the occurrence of false positives and false negatives, we may set up the anomaly detection task as one of estimating the likelihood that any given point is an anomaly, rather than classifying whether it is anomalous. Another frequently used perspective is to evaluate the relative anomalousness of different points: we may say that a point p is more anomalous than another point q, but may not be certain whether either is a true anomaly. To address this task, a possible scenario is to employ an algorithm to identify ten data points, rank-ordered, as being the most likely anomalous cases. These ten cases provide a starting point for human analysis or intervention, although all ten may turn out to be manifestations of acceptable behavior.

In the cybersecurity context, anomaly detection algorithms must account for the fact that the processes of interest are often neither deterministic nor completely random. Indeed, the difficulties encountered in cybersecurity applications can often be attributed to human actors with some free will. In other words, the observable behaviors are the result of the deliberate (non-random) actions of humans who are not predictable since their intentions are unknowable and their plans may change in unexpected ways. At best, making some observations over time can reveal a pattern that may be expected to continue, leading to some predictability. The human intent or plan causing that pattern may itself abruptly change, especially in response to cyber-defense mechanisms placed to thward previous plans, but such changes are not expected to occur very often.

---

[2]In medical diagnosis of tumors, for instance, it may be safer for preliminary analysis to permit more false positives than false negatives due to the much higher risk associated with missing a malignant tumor; however, an excess of false positives will render the analysis unsatisfactory.

2.1.1 Metrics for Measurement

To evaluate the performance of the algorithms, three metrics are often used: precision, recall, and Rank-Power [10, 17, 86, 103]; these are defined below.

Given a data set $\mathcal{D}$, suppose an outlier detection algorithm identifies $m > 0$ potential anomalies, of which $m_t (\leq m)$ are known to be true outliers. Then *Precision*, which measures the proportion of true outliers in top m suspicious instances, is:

$$Pr = \frac{m_t}{m},$$

and equals 1.0 if all the points identified by the algorithm are true outliers. If $\mathcal{D}$ contains $d_t (\geq m_t)$ true outliers, then another important measure is *Recall*, defined as:

$$Re = \frac{m_t}{d_t},$$

which equals 1.0 if all true outliers are discovered by the algorithm.

For the credit card transaction fraud example, if the data set contains 1,000,000 transactions of which 200 are fraudulent, an algorithm which considers all data to be anomalous exhibits $Pr = 0.0002$ and $Re = 1.0$, whereas an algorithm with 8 true positives (anomalies) and 2 false positives exhibits $Pr = 0.8$ and $Re = 10/200 = 0.05$.

Precision and recall are insufficient to capture completely the effectiveness of an algorithm, especially when comparing algorithms that result in different numbers of anomalies.

In particular, precision can take a small value just because m is large. One algorithm may identify an outlier as the most suspicious while another algorithm may identify it as the least suspicious. Yet the values for the above two measures remain the same. Ideally, an algorithm will be considered more effective if the true outliers occupy top positions and non-outliers are among the least suspicious instances. The "RankPower" metric [107] captures this notion. Formally, if R_i denotes the rank of the ith *true* outlier in the sorted list of most suspicious objects, then the RankPower is given by:

$$RP = \frac{m_t(m_t + 1)}{2 \sum_{i=1}^{m_t} R_i}.$$

Rank-Power takes maximum value 1 when all d_t true outliers are in top d_t positions. When algorithms with the same m are compared (e.g., each of them identifies 30 data points as outliers), the larger values of all three of these metrics imply better performance.

Example 2.1 Consider a dataset $\mathscr{D}$ of size $n = 50$ that contains exactly 5 anomalies. Suppose that an anomaly detection algorithm identifies $m = 10$ data points as anomalous, of which $m_t = 4$ are true anomalies. In addition, let the true anomalies in $\mathscr{D}$ occupy ranks equal to 1, 4, 5, and 8 in the sorted list of truly anomalous data points. Then,

$$Pr = \frac{4}{10} = 0.4, \quad Re = \frac{4}{5} = 0.8, \text{ and } RP = \frac{4 \times 5}{2(1 + 4 + 5 + 8)} = 0.56.$$

2.1.2 Old Problems vs. New Problems

Some anomalies may have been encountered previously in data, e.g., due to past attacks which were identified and not forgotten. Prior data analysis may then have identified *signatures* of patterns associated with such anomalies. For example, many viruses have been cataloged based on the effects they produce, or the occurrences of certain code fragments within the viruses. Rules have been formulated to detect such occurrences, and anti-malware software routinely applies such rules to help isolate potential malware. We then consider the analyst's problem as consisting of a classification task, identifying which data belongs to the "safe" category, and which data belongs to each known malware category. Learning algorithms (such as Support Vector Machines[29] and backpropagation-trained Neural Networks [85]) have been used to develop models that enable analysts to perform this classification task.

However, the classification approach can work only in detecting known problems of specific kinds, whereas the greatest damage in cyber-security applications is caused by unknown problems newly created by bad actors. Furthermore, there exist problems whose manifestations in the data do not admit simple categorization; hence ameliatory actions cannot be performed rapidly enough to prevent a catastrophe, e.g., when an employee gives his password to someone over the telephone, and the security of a sensitive system is compromised.

In such cases, *anomaly detection* procedures and algorithms are called for. The basic assumption remains that the observable data do reflect anomalies in the underlying process or behavior. However the corresponding changes in the data are not expected to follow previously known patterns. We seek to detect that something has gone wrong, even if we are unable to say exactly how or why that has happened; we may not have an explicit model, pattern or rule that describes the anomalies.

2.1.3 What Kind of Data?

Most data analysis begins with the assumption that some process exists, and that there are rules, guidelines, or physical principles governing the possible randomness in the data. Another assumption is that the characteristics of this process hold in the entire set of data, e.g., a sample containing only 10% of the data should be expected to have the same distribution as the entire data set. For data observed over time, the same trends or cycles are expected to be observed in any reasonably large[3] subset of the data, whether the subset is drawn from the early or late part of the time sequence.

The choice of the anomaly detection algorithm should depend on the nature of the process generating the anomaly. For example, when the relevant process is the occurrence of a malware infection, the observed data may be the values stored in certain memory locations, which vary from values expected in the absence of an infection.

Example 2.2 The "threads" currently running on a computer's processors may be observable. Some threads represent normal activity for a given time of day and for given users. But other threads may provide evidence that an unexpected computer program is currently executing, the "process" that resulted in the specific observable "data" (set of threads). Each "data point" is then a binary vector indicating which programs are active and result in observable threads. An anomaly corresponds to an unexpected set of programs being active, i.e., a bit vector substantially different from "expected" or normal cases. Even though the dimensionality of such a space is 2^n where n is the number of programs, the "normal" cases may correspond to a relatively small set of possibilities since only a few programs may be expected to be simultaneously active in most circumstances.

In more interesting cases, analysis needs to be performed over time, as illustrated by the following examples:

- The precise sequence in which certain processes are executed may be important to signify malware.
- The time sequence of Internet nodes over which a message is routed, signifying whether it is anomalous.
- We may consider behaviors of individuals: a single snapshot may not indicate anything wrong, but a series of observations of the same individual (over time) may indicate variations from the norm.

2.1.4 What's a Norm?

We have already used words such as "normal" quite often, without clear definitions. But *rigor* comes with the risk of *mortis*: circumscribing the permitted meaning of a term may limit flexibility in its usage, and exclude desirable variants of meaning. We seek the middle path, i.e., providing a rough but flexible definition.

Statisticians have long used the notions of (arithmetic) mean, median, and mode to capture the norms associated with distributions. Each of these is a single scalar

---

[3]The sample must be measured over a time period substantially larger than the length of the cycles in the data.

or multidimensional vector, and the distance of a point from the mean (or median or mode) has been used to assess the degree to which the point is "abnormal". Use of the Mahalanobis distance [84], with the standard deviation along each dimension as a normalizing factor, removes the potential confusion due to incommensurable dimensions. If the data distribution is normal, the Mahalanobis distance measure has an important property: it is the square root of the log likelihood of a point belonging to the distribution, i.e., an indication of how "anomalous" is the point relative to the data.

But with non-normal (and multi-modal) distributions, this straightforward interpretation no longer applies. Hence we seek to describe the "norm" as a set of points rather than a single point. For example, the norm may consist of a collection of "cluster centroids" or alternatively the boundaries of clusters. In order to consider a point to be "abnormal," it must be substantially distant from each of the points in the norm.

Also, if the data set is characterized by variations in density over clusters in space, such variations must also be accounted for in determining whether a point is abnormal. For instance, a larger distance may characterize an anomaly near a less dense cluster, whereas a smaller distance is reasonable for an anomaly near a more dense cluster. *Local density,* i.e., the number of data points in a unit (hyper)volume, then turns out to be a critical notion in identifying which points are more anomalous. Indeed, some anomaly detection algorithms rely on local density estimation and distance calculations that take local density into account.

The next section begins with very simple data, and we steadily work up our way through more complex cases.

2.2 Outliers in One-Dimensional Data

Quantitative data is often analyzed by computing the statistics describing its distribution. For clarity, we begin with single-dimensional quantitative data, i.e., each data point is a single number. Further, we begin with the simplest possible distributions, which have been well-studied:

- *Uniform Distribution:* When data is distributed uniformly over a finite range, the mean and standard deviation merely characterize the range of values. If the neighborhood of any data point is as richly populated as any other point, it can be argued that there are no anomalous data points, even though the extreme lower and upper limits of the range are relatively far from the mean. One possible indication of anomalous behavior could be that a small neighborhood contains substantially fewer or more data points than expected from a uniform distribution.
- *Normal Distribution:* When data is distributed normally, the density of points decreases substantially as we move away from the mean. About 0.1% of the points are more than 3σ (three standard deviations) away from the mean, and only about 5×10^{-8}% of the points are more than six standard deviations away from

the mean.[4] Hence it is often the case that a threshold (such as $3 \times \sigma$) is chosen, and points beyond that distance from the mean are declared to be anomalous. One contrary perspective is that the existence of some points far away from the mean is just a consequence of the fact that a variable is normally distributed. We may hence argue that a set of points (away from the mean) is anomalous if and only if their number is substantially higher than the number expected if the data were to be normally distributed, e.g., if 2% of the data points are found beyond the 3σ threshold.

- *Other Unimodal Distributions:* Many unimodal distributions are not normal, e.g., when there is a strict lower bound for the range of data values. Examples include log-normal and Gamma distributions. As with the normal distribution, if the nature and characteristics of the distribution are known, one may seek to find thresholds beyond which a relatively small number (e.g., 1%) of the data points are found. We may again argue that a collection of points (in a small region of the data space) is anomalous if their number is larger than predicted by the statistics of the distribution.
- *Multimodal Distributions:* The distributions for some data sets have multiple modes, discovered only when the data is closely examined. Heuristics such as the 3σ rule are not useful with such distributions. Instead it is more useful to think of the data as consisting of a collection of *clusters* of data points.

Clustering plays an important role in anomaly detection. Points which do not belong to any cluster are candidates to be considered anomalous. In particular, points which are distant from neighboring clusters are the logical anomaly choices for any anomaly detection algorithm. What remains is the clarification of when a collection of points should be considered a cluster, and what it means for a point to be considered sufficiently distant from a cluster (or multiple clusters).

The informal definition of a *cluster* is that it is a collection of points that are near each other—leaving open the task of defining "near."

1. Density-based cluster identifications are very popular in identifying anomalous data. If the relative number of points (per unit distance) is substantially higher in a small region than the entire data set, the points in that region can be considered as a cluster. This is still not a strictly mathematical definition, since the phrases "small region" and "substantially higher" are inherently fuzzy. The distribution of densities can itself be analyzed, and (if unimodal) we can identify the density threshold beyond which the density is high enough to consider a region to contain a cluster. Note that the same data set may contain one region of higher density and another region of lower density that may also be considered to be a cluster.
2. Another perspective is to compare intra-group distances with inter-group distances, and argue that points form a cluster if they are substantially closer to each other (on average) than they are to the nearest points outside the cluster.

---

[4]This is the origin of the "Six Sigma" principles, although the usually stated goal of 0.00034% defect rate corresponds to 4.5σ, not 6σ.

Unfortunately, there are some data distributions in which this definition leads to trivial cluster identifications (e.g., placing almost all the points in the same cluster).

With either of the above definitions, note that some points may lie outside clusters. Some clustering algorithms allocate every point to some cluster, which is not necessary. Other algorithms assume that the number of clusters is fixed (and predetermined or provided by the user); this is again not necessary.

When clustering-based approaches are used for anomaly detection, points inside clusters of a minimal size are usually not considered to be anomalous. This "minimal size" is again an externally specified parameter, such as a threshold based on the distribution of sizes of clusters in the data set.

Not all points outside clusters need to be considered equally anomalous. One point may be considered "more anomalous" than another if it is farther away from the nearest cluster—where this distance itself may be evaluated in different ways (e.g., distance to the nearest point in the cluster, or distance to the centroid of the cluster).

2.3   Outliers in Multidimensional Data

The ideas described earlier may be extended to multidimensional data; however, increasing dimensionality results in additional complications. The fundamental issues involve the choice of the distance measure: the Euclidean distance measure may not be appropriate since it may be combining numbers with different dimensionality, but the choice of a common scaling factor is not obvious for all dimensions. In attempting to evaluate which of two points (outside all clusters) is closest to a cluster, for instance, the distance measure chosen may determine the winner. Normalization, e.g., linearly scaling data values along different dimensions to lie in the same range, is often needed, but raises the question of what kind of normalization is best; extreme values may distort the results of simple normalization approaches.

An interesting complication arises when one of the relevant dimensions (or attributes) is time, which needs to be given special status. In particular, "time series" data involve three attributes: one is a label (identifying an object), another indicates (absolute) time, and the third is a numeric value. Time values come from a fixed and discrete range, and the usual assumption is that data points exist for each label and each permissible time point.

Two kinds of problems arise in this context:

1. If only one time series exists, i.e., all data points carry the same "label," then a relevant question is whether one portion of the time series differs substantially from the rest.
2. Multiple time series may exist, and we need to determine which time series is substantially different from most of the others. For example, in practical

applications that arise in financial applications, we are often interested in learning whether one time series changes values in lock-step with other time series, even if the actual data values are different. If a time series does not change in such lock-step manner with other time series, it could be considered anomalous. If the data values refer to stock/share prices, the number of outstanding shares of two companies may differ, so that two stocks may have similar variations in time even though the prices at any given instant may differ substantially. Algorithms have to be devised to determine whether this is the case, perhaps after linearly normalizing all stock prices to be in the same range.

In evaluating anomalies in behaviors of individuals or systems, three kinds of comparisons can be performed:

1. How does the behavior of the individual at a given time (or during a specific time period) compare with his own behavior in the past? For example, does one's credit card record for one month show purchases of a substantially higher magnitude or purchases of a different category than the past?
2. How does one's behavior compare with the behavior of all the other individuals for whom data is available? For instance, do an auto salesman's sales figures continue to rise even when almost all auto sales figures in the country are decreasing?
3. How does the behavior of one member of a cluster or sub-population of individuals compare with that of other members in the same cluster?[5] For instance, are the sales figures for a Toyota salesman in Syracuse uncorrelated with those of other Toyota salesmen in Syracuse?

2.4 Anomaly Detection Approaches

This section presents the main principles underlying different algorithms and approaches used for anomaly detection, particularly in the context of cyber-security. The primary approaches can be characterized as distance-based, density-based, and rank-based.

- *Distance-based*: Points that are farther from others are considered more anomalous.
- *Density-based*: Points that are in relatively low density regions are considered more anomalous.
- *Rank-based*: The most anomalous points are those whose nearest neighbors have others as nearest neighbors.

---

[5]The usual expectation is that the observable data variables that describe the "behavior" are not used in the clustering step.

For each of these approaches, the nature of the data may be *supervised, semi-supervised,* or *unsupervised.*

- In the supervised case, classification labels are known for a set of "training" data, and all comparisons and distances are with respect to such training data.
- In the unsupervised case, no such labels are known, so distances and comparisons are with respect to the entire data set.
- In semi-supervised problems, labels are known for some data, but not for most others. For instance, a few cases of malware of a certain new category may be available, and a semi-supervised learning algorithm may attempt to determine which other suspected cases of malware belong to the same category. Algorithms often proceed in multiple phases, with the early phase assigning tentative labels to unlabeled data.

An unsupervised anomaly detection algorithm should meet the following characteristics:

1. Normal behaviors have to be dynamically defined. No prior training data set or reference data set for normal behavior is needed.
2. Outliers must be detected effectively even if the distribution of data is unknown.
3. The algorithm should be adaptable to different domain characteristics; it should be applicable or modifiable for outlier detection in different domains, without requiring substantial domain knowledge.

2.5   Evaluation Criteria

Every anomaly detection problem appears to have different characteristics, and an algorithm that performs well on one problem may not perform well on another. Indeed, it is difficult to characterize what it means to "perform well," and three questions can be posed in this context:

1. Can quantitative metrics be devised so that we can unambiguously say which of two data points in a given data set is "more anomalous"—without appealing to human intuition? An answer is required in order to minimize the number of false positives generated by anomaly detection algorithms, particularly in the unsupervised and semi-supervised contexts.
2. Each anomaly detection algorithm answers the above question in a procedural way, e.g., based on distance to k nearest neighbors. Can this implicit choice be justified on mathematical or rational grounds?
3. In some cases, algorithms also appeal to the desire to compute the results in a "reasonable" amount of time, ruling out the search for optimal solutions. In such cases, can we say anything about the quality of the obtained solutions when compared to the optimal solutions?

Information theory provides a possible answer to these questions. A high-level basis for this approach is that many real-life processes are amenable to succinct descriptions of their essence.[6] Hence a variation from such a succinct description of the process should be considered an anomaly.

Example 2.3 Perhaps a quadratic curve can be fitted through ten points with a small threshold of tolerance (i.e., permitted distance from the curve). But a fourth degree curve may be needed to fit twelve points that include these ten; we may hence argue that the two additional points require multiple additional parameters, hence a longer description. Those two points may then be considered anomalous.

Multimodal distributions with ellipsoidal peaks capture the characteristics of many realistic problems; the model learning step would be to obtain the parameters of such a distribution, e.g., the number, locations, and standard deviations (along different dimensions) for the peaks, using a learning algorithm based on *Mixture of Densities, Kernel Regression, Support Vector Machines*, or *Radial Basis Function* approaches. We may then consider the extent to which each data point varies from the expectations of such a model.

For the same data set, multiple models are possible, and the ones with greater model complexity may result in smaller error, at the potential cost of *over-fitting* and poor generalization (to new data for the same problem). Researchers in machine learning and statistics have addressed this tradeoff in many ways, e.g., by *regularization*, searching for a model that optimizes a linear combination of model complexity and system error. Another "minimal description length" (MDL) [100] perspective is to minimize the total number of bits needed to describe the model as well as the deviations of data from the model. The related "Akaike Information Criterion" (AIC) [6, 61] minimizes the sum of the number of model parameters k and the logarithm of the maximized likelihood function for the model; an additional correction factor of $k(k+1)/(n-k-1)$ is applied to account for finite sample size n. AIC and MDL can be shown to be very similar, except that the latter multiplies the model complexity penalty by an extra term.

Information content can then be obtained by comparing the actual data to the (multimodal) distribution believed to characterize most of the data, obtained using a learning algorithm as mentioned above. Given a data set $\mathscr{D}$ characterized by a (learned or known a priori) distribution D, we may argue that a point p_1 is more anomalous than a point p_2 if $(\mathscr{D} - \{p_2\})$ has greater deviation from D than does $(\mathscr{D} - \{p_1\})$.

Alternatively, for each point p in the data set $\mathscr{D}$, we may evaluate the probability $P_D(p)$ with which it is drawn from that distribution D. We then say p_1 is more anomalous than p_2 if $P_D(p_1) < P_D(p_2)$.

It is also useful to have an absolute notion of anomalousness, in addition to the relative one above. Given a probability threshold θ, we may say that a point p is anomalous if $P_D(p) < \theta$. Results would depend on the user-determined parameter

---

[6]*Understanding is compression, comprehension is compression!*—G. Chaitin [18].

θ: a relatively high value of θ (e.g., 0.1) may lead to the conclusion that too many points are anomalous, whereas an extremely low value of θ (e.g., 0.00001) would suggest that almost none of the points are anomalous.

The relative and absolute notions mentioned above can be combined, by applying a squashing function g to P_D, such that $g(P_D(p)) = 0$ if $P_D(p) < \theta$. When $P_D(p) > \theta$, we may use a nonlinear monotonically increasing function such as $g(P_D(p)) = (P_D(p) - \theta)^4$.

2.6 Conclusion

In this chapter, we have presented the key questions and principles of anomaly detection, beginning with the formulations of what it means for a data point to be anomalous, the different approaches to anomaly detection, and how we may evaluate and compare different anomaly detection algorithms. These considerations precede the detailed formulation of algorithms, and how we may tailor general-purpose algorithms to problem-specific criteria and constraints. Later chapters address these details.

Chapter 3
Distance-Based Anomaly Detection Approaches

In this chapter we consider anomaly detection based on distance (similarity) measures. Our approach is to explore various possible scenarios in which an anomaly may arise. To keep things simple, in most of the chapter we illustrate basic concepts using one-dimensional observations. Distance based algorithms, proposed by researchers, are presented in Chap. 6.

3.1 Introduction

By definition, identifying an anomaly involves figuring out that a data point is "different" from others. This definition is necessarily parameterized by the data set against which the data point is compared: a person who is five feet tall may be anomalous among male college basketball players, but not among horse-riding jockeys.

For convenience of presentation, throughout the chapter we assume that feature observations belong to continuous space, but the discussion also applies to discrete and nominal attributes where a suitable distance metric can be defined, see [14, 55]. In continuous spaces where all data attributes are real-valued (possibly within a bounded range), we would say a data point is "different" from others if its distance to other points is large. However, anomaly detection algorithms differ in how this distance is evaluated. This is because no consensus exists on which sets of points are to be used to compare distances, nor on how to evaluate the *distance to a collection of points*, even though most researchers agree to work with the standard and well-known definitions of the distance between two points.

This chapter focuses on anomaly detection algorithms that rely on distance calculations between points and sets of points. The symbol $\mathscr{D}$ is used in this text to represent the n-dimensional data set, presumed to be in continuous unbounded real-valued space $\mathcal{R}^n$. Data points in $\mathscr{D}$ are denoted by characters p, q, possibly

© Springer International Publishing AG 2017

K.G. Mehrotra et al., *Anomaly Detection Principles and Algorithms*, Terrorism, Security, and Computation, https://doi.org/10.1007/978-3-319-67526-8_3

carrying subscripts or superscripts. Uppercase symbols such as P are used to denote sets of points, i.e., $P \subset \mathscr{D}$. The distance between two points $p, q \in \mathscr{D}$ is denoted by $d(p, q)$.

We begin with the simplest algorithms, giving examples and counter-examples to indicate the usefulness and limitations of such algorithms. We address three primary questions:

1. **Measurement:** How anomalous is a given point? This requires transforming points into a one-dimensional scale, e.g., defining a function α such that $\alpha(p) \in \mathcal{R}$ measures the anomalousness of $p \in \mathscr{D}$.
2. **Absolute:** Is a given point anomalous? This requires finding a threshold $\theta > 0$ so that we say a point $p \in \mathscr{D}$ is anomalous if $\alpha(p) > \theta$.
3. **Relative:** Is one point "more anomalous" than another? This requires comparing points, so that $\alpha(p) > \alpha(q)$ if a point $p \in \mathscr{D}$ is more anomalous than another point $q \in \mathscr{D}$. We may denote this relationship with the symbol "$\triangleright$", i.e., $p \triangleright q$ if $\alpha(p) > \alpha(q)$.

Two secondary questions are also posed in practice, and their answers can be derived from the former:

- **Subset:** What are the m most anomalous points in a given data set? To answer this question, we may use the "relative" criterion to find the most anomalous, the second most anomalous, etc., perhaps by sorting if m is large.
- **Hybrid:** What are the m most anomalous points in a given data set, which are also absolutely anomalous? The answer can be obtained by applying an absolute threshold to the "Subset" answer above.

In Sect. 3.2 we consider some of the distance (similarity) measures, in Sect. 3.3 we describe some distance based approaches, Sect. 3.4 contains the summary and comments.

3.2 Similarity Measures

Often, we define similarity in terms of a distance measure. Note that a similarity measure can be obtained by 'essentially' considering the 'inverse' of a distance measure.

Popularly applied similarity measures include direct measures such as the Euclidean, Minkowski, and Mahalanobis measures. Additional measures, such as cosine similarity and Jaccard index are also used; often indirect measures, such as Shared Nearest Neighbor (SNN) are more appropriate.[1]

---

[1] In SNN, the number of common points between k-nearest neighbors of two points represents the desired similarity between the points. This measure is described in more detail in Chap. 6.

When the different dimensions in the data set $\mathscr{D}$ are incommensurable and have different mutual correlations, a preferred measure is the Mahalanobis distance

$$\sqrt{(p-q)^T S^{-1}(p-q)}$$

where S is the covariance matrix measuring the mutual correlations between dimensions for all points in the data set $\mathscr{D}$. If the covariance matrix S is diagonal, this simplifies to

$$\sqrt{\left(\sum_{i=1}^{d}(p_i - q_i)^2/\sigma_i^2\right)}$$

where σ_i is the standard deviation in the ith dimension. In the simplest case, where S is the identity matrix, this reduces to the most commonly used Euclidean distance measure,

$$d(p,q) = \sqrt{\sum_{i=1}^{d}(p_i - q_i)^2}.$$

Often, data is first linearly normalized along each dimension so that every point lies in $[-1, +1]^d$, before Euclidean distance calculations are performed.

The Minkowski distance of order ℓ between two points $p = (p_1, \ldots, p_d)$ and $q = (q_1, \ldots, q_d) \in \mathscr{D}$ is defined as:

$$\left(\sum_{i=1}^{d}|p_i - q_i|^\ell\right)^{\frac{1}{\ell}}.$$

Most often used values of ℓ are 1 and 2; for $\ell = 2$ the Minkowski distance is equal to the Euclidean distance and for $\ell = 1$ this distance is equal to $\left(\sum_{i=1}^{d}|p_i - q_i|\right)$ and is known as the Manhattan distance. Two other special cases of Minkowski distance, defined below, are also used:

$$\max_{i=1}^{d}|p_i - q_i|$$

and

$$\min_{i=1}^{d}|p_i - q_i|.$$

Jacob similarity is defined for two sets of data points. Let A and B be two datasets, then

$$J(A, B) = \frac{|A \cap B|}{|A| + |B| - |A \cap B|}.$$

Jaccard similarity is defined for binary points p and q, i. e., when all coordinates of both take binary values only; i. e., $p_i = 0$ or 1, for $i = 1, \ldots d$.

$$J(p, q) = \frac{m_{11}}{m_{01} + m_{10} + m_{11}},$$

where m_{11} = number of places where p and q are both 1, m_{01} = number of places where p_i's are 0 and q_i's are 1, and m_{10} = number of places where p_i's are 1 and q_i's are 0. For example, let $p = (1, 1, 0, 0, 1, 0)$ and $q = (1, 0, 1, 0, 0, 1)$, then $m_{11} = 1, m_{01} = 2, m_{10} = 2$, and $J(p, q) = 1/5$.

Cosine similarity between two non-zero binary points $p, q \in \mathscr{D}$ is defined as

$$\frac{\sum_{i=1}^{d} p_i q_i}{\sqrt{\sum_{i=1}^{d} p_i^2 \sum_{i=1}^{d} q_i^2}}.$$

The resulting similarity ranges from -1 to 1; when this similarity is 1, $p = q$ and if the similarity is -1, then p and q are exact opposite of each other.

3.3   Distance-Based Approaches

In this section, we present several simple anomaly detection algorithms and approaches based on distance computations alone.

3.3.1   Distance to All Points

The simplest possible anomaly detection algorithm would evaluate each point $p \in \mathscr{D}$ against all points in $\mathscr{D}$. The sum of distances from all points can be used as the anomalousness metric, i.e.,

$$\alpha(p) = \sum_{q \in \mathscr{D}} d(p, q).$$

The most anomalous point is farthest from all points in the data set (Fig. 3.1).

Fig. 3.1 Dot plot of dataset $\mathscr{D} = \{1, 2, 3, 8, 20, 21\}$; *red dot* corresponds to anomalous observation

Fig. 3.2 Dot plot of dataset $\mathscr{D} = \{1, 3, 5, 7, 100, 101, 200, 202, 205, 208, 210, 212, 214\}$, *red dot* indicates anomalous observation

Example 3.1 Consider the one-dimensional data set $\mathscr{D} = \{1, 2, 3, 8, 20, 21\}$. $\alpha(1) = 0 + 1 + 2 + 7 + 19 + 20 = 49$. Likewise, the alpha values for the remaining points are 45, 43, 43, 67, and 71, respectively. Hence, according to this criterion, the most anomalous point is 21. This illustrates that extremal points are likely to be chosen as anomalous, by this approach, which is often not the desired goal.

3.3.2 Distance to Nearest Neighbor

In this simple approach, we focus on the distance to the nearest point in the data set, i.e.,

$$\alpha(p) = \min_{q \in \mathscr{D}, q \neq p} d(p, q).$$

The most anomalous point is one whose nearest neighbor is at the greatest distance (Fig. 3.2).

Example 3.2 Consider again the one-dimensional data set $\mathscr{D} = \{1, 2, 3, 8, 20, 21\}$. $\alpha(1) = \min(1, 2, 7, 19, 20) = 1$, and the alpha values for the remaining points are 1, 1, 5, 1, and 1, respectively. Hence the most anomalous point is 8, according to this criterion, which is usually considered satisfactory for such a data set.

Example 3.3 Consider the larger one-dimensional data set $\mathscr{D} = \{1, 3, 5, 7, 100, 101, 200, 202, 205, 208, 210, 212, 214\}$. In this case, the α values would be 2, 2, 2, 2, 1, 1, 2, 2, 3, 2, 2, and 2, respectively, suggesting that the most anomalous point is 205, although it is at the center of a relatively close collection of points. This, again, is often not the desired goal, and the problem can be attributed to the fact that only a single near neighbor is insufficient to identify a point as non-anomalous.

3.3.3 Average Distance to k Nearest Neighbors

This approach requires a parameter $k < N = |\mathscr{D}|$ that identifies the number of nearest neighbors to be considered. For convenience, we first define $Near(p, j)$ to be a jth nearest neighbor of a point $p \in \mathscr{D}$, where $j < N = |\mathscr{D}|$, breaking ties arbitrarily. Thus, the nearest neighbor of $p \in \mathscr{D}$ would be $Near(p, 1)$. The average distance to k nearest neighbors, used as the indicator of the anomalousness of a point p would be:

Fig. 3.3 Dot plot of dataset $\mathcal{D} = \{1, 3, 5, 7, 100, 101, 200, 202, 205, 208, 210, 212, 214\}$, *red dot indicates anomalous observation*

$$\alpha(p) = \sum_{j=1}^{k} d(p, Near(p, j))/k.$$

Equivalently, we may use the sum of the distances, omitting the division by k (Fig. 3.3).

Example 3.4 Consider again the one-dimensional data set $\mathcal{D} = \{1, 3, 5, 7, 100, 101, 200, 202, 205, 208, 210, 212, 214\}$. For $k = 3$, the α values (sums) would be 12, 8, 8, 12, 189, 191, 15, 11, 11, 9, 8, 8, and 12, respectively, suggesting that the most anomalous point is 101. For $k = 4$, the α values (sums) would be 111, 105, 103, 105, 286, 289, 25, 19, 16, 15, 13, 15, and 21, respectively, still suggesting that the most anomalous point is 101. The choice of k implicitly determines how many points must be very near to another in order for it to be considered non-anomalous. A very small value of k will result in false negatives, i.e., not flagging anomalous points, as happens when $k = 1$ and this approach becomes identical to the criterion of the distance to the single nearest neighbor (mentioned in the preceding subsection). Conversely, a large value of k may result in non-intuitive values, e.g., $k = |\mathcal{D}| - 1$ corresponds to the first case in this section, using the sum of the distances to all points in the data set, so that extremal values are flagged as most anomalous.

3.3.4   Median Distance to k Nearest Neighbors

The arithmetic average is not very robust in that the addition of one or two more points may drastically change the outcome of the computation. We may instead use the median, a more robust measure, less sensitive to noise in the data, although it requires a greater amount of computation. If $k = 2m - 1$ is odd, the median distance to k nearest neighbors is the same as the distance to the mth nearest neighbor.

Example 3.5 Consider again the one-dimensional data set $\mathcal{D} = \{1, 3, 5, 7, 100, 101, 200, 202, 208, 210, 212, 214, \}$. For $k = 3$, the α values would be 4, 2, 2, 4, 93, 94, 5, 3, 3, 3, 2, 2, and 4, respectively, suggesting that the most anomalous point is 101. As before, the results of the computation are sensitive to the choice of k.

3.4 Conclusion

This chapter has presented a collection of simple anomaly detection algorithms that are based on distance computations alone. The following chapters explore the use of other concepts, such as clusters, providing more elaborate algorithms and approaches used in practice by many researchers.

Chapter 4
Clustering-Based Anomaly Detection Approaches

This chapter explores anomaly detection approaches based on explicit identification of clusters in a data set. Points that are not within a cluster become candidates to be considered anomalies. Variations among algorithms result in evaluating the relative anomalousness of points that are near (but not inside) a cluster, and also the points at the periphery of a cluster.

In Sect. 4.1, we describe some classical clustering algorithms such as the k-means clustering, followed by how to discover asymmetric clusters; that is, a procedure in which a cluster is dynamically obtained, typically one at a time. Finally, Sect. 4.2 discusses basic concepts of anomaly detection using clusters.

4.1 Identifying Clusters

Clustering can be based on *similarity* or *distance* computations; these two approaches differ, although the end result is often the same because similarity measures are strongly negatively correlated with distance measures. Distance-based clustering approaches are based on the idea that points within the same cluster are separated by relatively small distances, whereas points in different clusters are at greater distance from each other. Similarity-based clustering approaches suggest that points that are similar to each other should belong in the same cluster because points at smaller distances from each other are presumed to be more similar.

Clustering algorithms generally presume that the data is in a bounded continuous multi-dimensional space, and that a similarity or distance measure has already been chosen. In general, each point p_i is assigned a "degree of membership" $\mu(p_i, C_j) \in [0, 1]$ to any cluster C_j. A few variations exist, such as the following:

- Most clustering algorithms partition the data into clusters, i.e., place every point in some cluster, so that each $\mu(p_i, C_j) \in \{0, 1\}$ and $\sum_j \mu(p_i, C_j) = 1$ for each p_i.

© Springer International Publishing AG 2017
K.G. Mehrotra et al., *Anomaly Detection Principles and Algorithms*, Terrorism, Security, and Computation, https://doi.org/10.1007/978-3-319-67526-8_4

- Non-partitioning algorithms allow for the possibility that some points do not belong to any cluster, i.e., each $\mu(p_i, C_j) \in \{0, 1\}$ and $\sum_j \mu(p_i, C_j) \leq 1$ for each p_i.
- Some algorithms allow clusters to overlap, and a point may belong to multiple clusters, i.e., each $\mu(p_i, C_j) \in \{0, 1\}$ and $\sum_j \mu(p_i, C_j) \leq$ the number of clusters, for each p_i.
- *Fuzzy*[1] clustering algorithms permit non-binary membership values, i.e., $0.0 \leq \mu(p_i, C_j) \leq 1.0$, although they often restrict $\sum_j \mu(p_i, C_j) = 1$ for each p_i.

We present a few well-known algorithms for identifying clusters; some that place every point in some cluster, some that allow for the possibility that a point may not belong to any cluster. More exhaustive coverage of clustering methods is provided in textbooks such as [5, 37, 62].

4.1.1   Nearest Neighbor Clustering

k-Nearest Neighbor algorithms have been proposed for many kinds of problems, based on the main idea that an individual should be similar to a majority of its k immediate neighbors, rather than to a centroid or an aggregate over a large set of data points. This approach hence gives greater weightage to local properties of data spaces, although computationally more expensive and contrary to the philosophy of statistics to summarize properties of large amounts of data into a few numerical characteristics. This approach has been used for classification tasks [34], in the context of supervised learning, and can also be applied to clustering tasks if a set of points distant from each other are initially chosen to be labeled with different cluster-IDs, analogous to class labels in a classification problem. Points are labeled with the same label as a majority of their k nearest neighbors.

A "Region-growing" heuristic is to start with a single new point at a time, label it with a new cluster-id, and iteratively label all the immediate neighbors of labeled points whose distance (to a labeled point) is smaller than a threshold that depends on the distances between labeled points in that cluster. At the conclusion of this step, "relaxation" iterations may be carried out, labeling points based on the majority of their k immediate neighbors; these iterations may start at the boundaries of the previously identified clusters, and result in merging some clusters.

---

[1]Fuzzy membership allows for non-binary cluster allocation, and is not the same as probability of membership. Saying that a point belongs to a cluster with a certain probability, assumes that there are only two cases: either the point belongs, or does not belong, to that cluster; but there is uncertainty regarding which case holds. The discovery of future information or evidence may reveal whether or not the point belongs to the cluster. On the other hand, when we say $\mu(p_i, C_j) = 0.7$, this is not a probability, but may instead express the fact that p_i is near the other points with high membership values in C_j, but other points with membership values > 0.7 are even nearer.

Single nearest neighbor clustering has been proposed earlier in the context of agglomerative clustering, e.g., the SLINK algorithm [104] which constructs a dendrogram with $O(n^2)$ computational effort. In such an algorithm, centroid computation is not needed, and distance computations are performed with respect to individual data points. Each node in the hierarchy is explicitly associated with a set of points, and not just the centroid, so that there is no assumption of clusters being symmetric in any dimension. Two clusters C_i, C_j are combined if they contain points ($x \in C_i$, $y \in C_j$) whose Euclidean distance $d_E(x, y) < \theta$ is considered small enough, even if other points in the clusters are far away from each other. The threshold θ may be a function of the data set, e.g., (mean + $3 \times \sigma$) of the distances in the minimum spanning tree for the data set.

4.1.2   k-Means Clustering

This well-known algorithm [81], described in Algorithm "k-means clustering", repeatedly computes the centroid of each "current" cluster, and then updates the same after re-associating each data point with the nearest (current) cluster centroid.

Algorithm k-means clustering

Require: Data set $\mathcal{D}$, Number of clusters (k), termination threshold θ;
Ensure: k Clusters
1: **Initialize:** Randomly select k distinct elements of $\mathcal{D}$ as the initial set of centroids $C = \{c_1, \ldots, c_k\}$;
2: **Repeat**
3:                    Assign each $p \in \mathcal{D}$ to the closest cluster, minimizing $d_E(p, c_j)$;
4:                    **Update** each c_j to be the centroid (mean) of the points assigned to it;
5:   **Until** the number of points reassigned in this iteration $< \theta$;

This algorithm implicitly identifies each cluster with its cluster centroid, resulting in symmetric clusters such as spheres for three-dimensional data. Although this is perhaps the most widely used clustering algorithm, there are many problems for which the k-means clustering approach is unsatisfactory, as illustrated in Fig. 4.1.

Another difficulty with this algorithm is that the initial choice of the starting points can affect the final result.

Example 4.1 Let $\mathcal{D} = \{1, 4, 6, 10\}$ and let $k = 2$.

- If the initially chosen cluster centroids are 1 and 10, respectively, the first iteration of the k-means clustering algorithm would result in the clusters $\{1, 4\}$ and $\{6, 10\}$, since 4 is closer to 1 (than 10) and 6 is closer to 10 (than 1).
- But if the initially chosen cluster centroids are 1 and 4, respectively, the algorithm instead yields the clusters $\{1\}$ and $\{4, 6, 10\}$, whose centroids are 1 and 6.66, respectively.

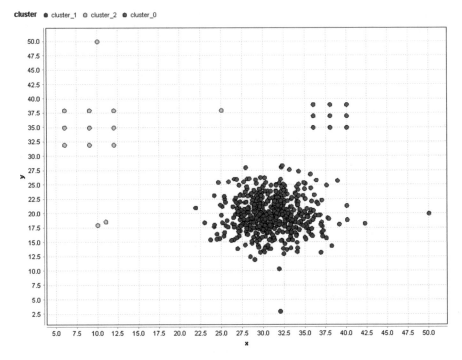

Fig. 4.1 Asymmetric clusters and the results of applying k-means algorithm with $k = 3$; clusters are identified by the colors of the points

- No further change occurs, in each case, since the two cluster centroids (2.5 and 9.5 in the first case; 1 and 6.66 in the second case) are closest to all the points in the respective clusters they represent.

An additional difficulty with this algorithm is the determination of the number of clusters (k). One guideline is that the ratio of the intra-cluster distance to the inter-cluster distance should be small. Some implementations begin with small k and successively increment it as long as significant measurable progress is achieved, e.g., in terms of the above ratio or the *silhouette* measure

$$s = 1 - (\text{distance to own centroid})/(\text{distance to next nearest centroid}).$$

In the "elbow" method, the number of clusters are chosen at the point where the incremental improvement is small.

Information-theoretic criteria have also been formulated to determine the best choice of k. For each data point $q \in \mathscr{D}$, the maximum *Likelihood* function $L(\Theta|q)$ is proportional to the probability $P(q|\Theta)$, where Θ is the best Gaussian mixture distribution described by the result of the clustering algorithm. The overall maximum likelihood that the available data corresponds to the Mixture distribution is then $L = P(\mathscr{D}|\Theta)$. The *Akaike Information Criterion* suggests that we choose

k to minimize $k - \ln(L)$, whereas the *Bayesian Information Criterion* minimizes $-\ln(L)$. Another variant is based on *rate-distortion theory*, with an analytical expression used to determine the amount of data compression that can be achieved. Unfortunately, the latter approaches are computationally expensive and are hence impractical for large data sets.

4.1.3 Fuzzy Clustering

The k-means clustering algorithm is expected to minimize $\sum_{p_i} \sum_{c_j} \|p_i - c_j\|^2$, where each $p_i \in \mathcal{D}$ and each c_j represents centroid of the jth cluster. However, it may be argued that points that are closer to a centroid should be given greater weightage in the minimization process, and that the cluster allocation should depend less on points that are distant from cluster centroids. If $\mu_{i,j} \in [0, 1]$ represents the degree to which p_i belongs to the fuzzy cluster whose weighted centroid is c_j, estimated as a decreasing function of the distance between p_i and c_j, then Bezdek's "Fuzzy k-means" clustering algorithm [12] attempts to minimize

$$\sum_{p_i} \sum_{c_j} \mu_{i,j}^m \|p_i - c_j\|^2,$$

generalizing the k-means clustering algorithm in an intuitive way, as shown in Algorithm "Fuzzy k-means clustering".

Algorithm Fuzzy k-means clustering

Require: Data set $\mathcal{D}$, Number of clusters (k), termination threshold θ;
Ensure: k Clusters
1: **Initialize:** Randomly select k distinct elements of $\mathcal{D}$ as the initial set of centroids $C = \{c_1, \ldots, c_k\}$;
2: **Repeat**
3: Compute the membership of each $p \in \mathcal{D}$ in each cluster to be a decreasing function of $d_E(p, c_j)$, such as $1/exp(d_E(p, c_j)^2)$ or $1/(d_E(p, c_j))^m$ (e.g., with $m > 2$);
    Normalize membership values so that $\sum_j \mu(p, c_j) = 1$;
4: **Update** each c_j to be the membership-weighted centroid of the points assigned to it,

$$c_j = \frac{\sum_i \mu_{i,j}^m p_i}{\sum_i \mu_{i,j}^m};$$

5: **Until** the magnitude of the changes to fuzzy cluster centroids in this iteration $< \theta$;

As in the k-means algorithm, the membership degrees and weighted centroid positions are iteratively updated until convergence, i.e., when little further change is observed, using a predetermined threshold. As with most iterative minimization algorithms, there is no guarantee that a global minimum will be reached for the function being optimized. At a high level, this algorithm is similar to the well-known expectation-maximization algorithm [32], wherein an explicit assumption is made that the data consists of a mixture of Gaussian distributions.

4.1.4 Agglomerative Clustering

This approach, described in Algorithm "Agglomerative clustering", begins with many tiny clusters, each containing a single element, and successively merges small clusters to form bigger clusters. Two clusters are candidates for merging if they are nearest to each other, e.g., based on the distance between cluster centroids. The process can be terminated when the number or sizes of clusters are considered satisfactory, or until the next merger would result in a single cluster.

Algorithm Agglomerative clustering

Ensure: Data set $\mathscr{D}$
1: Initialize: Each data point is in a cluster by itself, constituting the current frontier of the dendrogram;
2: **Repeat**
3: Find two nodes (in the current frontier) with the shortest distance between their cluster centroids;
4: Merge them, forming their parent node in the dendrogram;
5: Replace the two pre-merger nodes by the new merged node in the current frontier, along with its newly computed centroid;
6: Compute its distances from other nodes' centroids in the frontier;
7: **Until** all nodes have been merged (or computational resources are exhausted).

This algorithm does not require an externally determined parameter for the number of clusters (k), and is deterministic, giving the same result in each execution, unlike the k-means algorithm. But it requires more computational effort, and external decision-making may still be required to determine how deep we need to go along each path down from the root before declaring a node as representing a genuine cluster.

Sibson's *Single Linkage (SLINK)* [104] algorithm reduces the complexity from $O(n^3)$ to $O(n^2)$, and is based on nearest-neighbor clustering rather than centroid computation.

The merging process is essentially binary: at each step, two nodes come together; in some problems, a tree with branching factor more than two may be more representative of the data. Variations of the algorithm can be defined to permit multiple nodes to merge in a single iteration.

BIRCH (Balanced Iterative Reducing and Clustering using Hierarchies) [120] and CURE (Clustering Using REpresentatives) [50] are efficient agglomerative clustering techniques that facilitate anomaly detection.

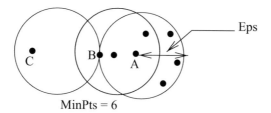

MinPts = 6

Fig. 4.2 Point A is a core point because its neighbor contains Eps = 6 points; Point B is a border point because it belongs to A's neighborhood but its neighborhood does not contain Eps points; point C is a noise/anomalous point

4.1.5 Density-Based Agglomerative Clustering

Some applications require the rapid discovery of asymmetric clusters; this has been addressed in the Density-Based Spatial Clustering of Applications with Noise (DBSCAN) algorithm [36], intended to discover clusters of arbitrary shape from noisy data sets, requiring only one scan of the dataset.

The central notion underlying this agglomerative algorithm is that of a *core point*, in whose vicinity (within a distance of r) lie a minimal number of other points $\geq \theta_\#$) in the data set. Initially, clusters are formed around core points; in each iteration, two clusters are merged if they contain core points within a distance of r from each other. Non-core points in such clusters are called *border points*, whereas points outside such clusters are called *noise points*, i.e., outliers. Figure 4.2 illustrates these three types of points for a small dataset. In this figure A is a core point because its neighborhood contains MinPts = 6 and B is a border point. Note that B is in the neighborhood of A but the neighborhood of B contains 4 < MinPts = 6. C is a noise point because it is neither a border point nor a core point.

Algorithm DBSCAN

Ensure: Data set $\mathcal{D}$, threshold parameters r, $\theta_\#$
1: For each point $p \in \mathcal{D}$,
2:    Find its neighbors $q \in N(p) \subseteq \mathcal{D}$ such that $d_E(p, q) < r$;
3:    Label p as a core point $\in C$, the initial frontier, if $|N(p)| \leq \theta_\#$;
4: Endfor;
5: For each point $p \in \mathcal{D}$,
6:        add p to the cluster associated with some $c \in C$ if $d_E(p, c) < r$;
7:
8: **Repeat**
9:
10:    If the current frontier contains two nodes c_1, c_2 such that $d_E(c_1, c_2) < r$, then
11:    Merge the clusters they represent, and form their parent node in the dendrogram;
12:    Replace the two pre-merger nodes by the new merged node in the current frontier, along with its newly computed centroid;
13: **Until** no more mergers are possible.

DBSCAN has problems in clustering some datasets that consist of clusters with widely varying densities. We note that DBSCAN uses the traditional concept of density; defined as the number of points in unit volume. This definition falls apart in case the features are discrete or categorical, and the dimensionality of feature vector is high; in such cases, the distribution of Euclidean distance between points tends to be uniform. This deficiency has been addressed by several researchers including Ertöz et al. [35], Jarvis and Patrick [63], and Guha et al. [51] by generalization of the concept of density and/or by using other measures of similarity such as *links*, Jaccard measure or cosine measure. The measure of similarity between two points p and q that counts the number of common neighbors in sets of their k-nearest neighbor is particularly significant when the feature variables are discrete/categorical and when the dimensionality of the feature vector is large.

CURE [50] is another agglomerative hierarchical algorithm that represents each cluster using a set of *representative points,* away from the centroid. Two clusters can be merged if the minimum distance between any two of their representative points is $<$ a threshold. CURE identifies outliers by considering the growth of a cluster; a slow growing cluster implies that the associated observations are likely to be anomalies. The first representative point is chosen to be the point furthest away from the center of the cluster and subsequent point is chosen so that it is farthest from all the previously chosen points, This guarantees that representative points are well distributed. Once the representative points are chosen, they are shrunk toward the center by a factor which helps to moderate the effect of noise points located at cluster boundaries (the representative points belonging to the clusters and noise points are thus farther apart).

In Chap. 7, we consider another generalization of DBSCAN, using ranks and associated anomaly detection algorithms.

4.1.6 Divisive Clustering

In this approach, a dendrogram is instead produced by successively partitioning sets of data points. The process may be permitted to terminate when each cluster is:

- Small: the number of elements in each cluster $\leq$ a threshold; or
- Tight: the maximum distance between data points in a cluster $\leq$ a threshold;
- Relatively compact: average distance within the cluster is much smaller than in the entire data set.

Avoiding the extensive distance computations required at the lowest (fine-grain) levels imply that this algorithm would be more computationally efficient than agglomerative clustering.

Algorithm Divisive clustering
Ensure: Data set $\mathscr{D}$
1: Initialize: The dendrogram (and the current frontier) contains a single splittable node representing the entire data set $\mathscr{D}$;
2: **Repeat**
3: Split each splittable node in the current frontier into two nodes, e.g., using k-means clustering with $k = 2$;
4: Identify and mark the "trivial" nodes in the frontier, on which further splitting need not be performed;
5: **Until** all nodes have been marked as trivial.

4.2 Anomaly Detection Using Clusters

Assuming that clusters have been identified using algorithms such as those defined in the previous section, we now address the task of identifying anomalous points. Several possible approaches are defined in the rest of this section.

4.2.1 Cluster Membership or Size

As mentioned earlier, some clustering algorithms permit data points to lie outside identified clusters and in such cases, the points that do not belong to any cluster can be considered anomalous. A simple example illustrates this concept.

Example 4.2 Consider the one-dimensional data set $\mathscr{D}_1 = \{1, 2, 3, 4, 5, 30, 31, 51, 52, 53, 55, 80, 110, 111, 112, 113\}$. If a clustering algorithm identifies the three clusters $\{1, 2, 3, 4, 5\}$, $\{51, 52, 53, 55\}$, and $\{110, 111, 112,113\}$, then the data points 30, 31 and 80, which lie outside the clusters, are considered to be the anomalous cases.

   The outcome of this approach depends on the predetermined thresholds, such as the number of clusters and minimum number of points required in a cluster. The following examples illustrate how decisions are effected by the thresholds.

Example 4.3 Consider the one dimensional data set $\mathscr{D}_1 = \{1, 2, 3, 4, 5, 30, 31, 51, 52, 53, 55, 80, 110, 111, 112,113\}$ of the previous example. If the k-means clustering algorithm is applied for $k = 3$ to this data, the three clusters identified would all be sufficiently large so that no data points would be flagged as anomalous. But if $k = 5$, we may obtain two clusters, $\{30, 31\}$ and $\{80\}$, that contain a relatively small number of data points; hence points therein are anomalous.

- If a cluster size threshold of 3 is used, then the elements 30, 31, and 80 would be considered anomalous.
- If the cluster size threshold had been 2, then only the element 80 would be considered anomalous.

Some algorithms such as DBSCAN explicitly enforce a minimum size requirement for a cluster; but it is not obvious how to choose this threshold.

Instead of an absolute threshold (such as 3), the threshold for large data sets may be a function of the size of the data set $|\mathscr{D}|$ and the number of clusters k, e.g., the threshold may be $|\mathscr{D}|/10k$, following the rationale that the expected size of each cluster is $|\mathscr{D}|/k$, and no real cluster should be smaller by an order of magnitude.

4.2.2 Proximity to Other Points

A partitioning algorithm, such as the k-means clustering, places every point in some cluster and every such cluster may satisfy the size condition; that is, the size of each cluster is $\geq$ the size threshold. But some points in the cluster may be so far from all others in the same cluster that we may identify them as anomalous.

Example 4.4 Consider the one-dimensional data set used earlier: $\mathscr{D}_1 = \{1, 2, 3, 4, 5, 30, 31, 51, 52, 53, 55, 80, 110, 111, 112, 113\}$ to which the k-means clustering algorithm is applied. If $k = 3$, we identify three clusters $\{1, 2, 3, 4, 5\}$, $\{30, 31, 51, 52, 53, 55\}$, and $\{80, 110, 111, 112, 113\}$. In the second cluster, each of the points 30 and 31 is substantially farther from other points than are 51, 52, 53, 55; this suggests that these points are anomalous. Likewise, using a similar argument point 80, in the third cluster, is anomalous.

Since calculating the sum of distances to all points in a cluster is computationally expensive, a surrogate is to compute the distance of each point from the centroid of the cluster. This approach can also be applied with non-partitioning algorithms, i.e., when some points do not belong to any cluster. We then define

$$\alpha(p) = \min_j d(p, \mu_j),$$

where μ_j is the centroid of cluster C_j, i.e.,

$$\mu_j = \sum_{p_j \in C_j} p_j / |C_j|.$$

If $\alpha(p)$ is 'large' p_i is considered to be an anomaly.

Example 4.5 For the example considered above, with $k = 3$ clusters $C_1 = \{1, 2, 3, 4, 5\}$, $C_2 = \{30, 31, 51, 52, 53, 55\}$, and $C_3 = \{80, 110, 111, 112, 113\}$, are obtained and we find that $\mu_1 = 3, \mu_2 = 45.3, \mu_3 = 105.2$; and $\alpha(1) = 2$, $\alpha(30) = d(30, 45.3) = 15.3$ and $\alpha(80) = 25.2$, suggesting that the data point 80 is much more anomalous than 1 or 30. Since $\alpha(80) = 25.2$ and $\alpha(113) = 7.8$, we would consider 80 to be the more anomalous point then 113. Among all points considered 80 will be most anomalous point.

For this approach to work well, the number of clusters k must be correctly chosen, neither k can be very large, nor very small. If k is large, to some extent the difficulty may be overcome if a cluster size threshold is applied to the result. When k is very small, we can evaluate whether the intra-cluster distances are too large and if a small increase in k can substantially decrease the intra-cluster distances, then such an increase should be permitted. However, this is the essence of choosing the 'right' number of clusters in k-means clustering, discussed in Sect. 4.1.2.

4.2.3 Proximity to Nearest Neighbor

Algorithms that rely on the distances to cluster centroids have an implicit assumption that clusters should be "symmetric," e.g., circular in two dimensions, since two points at the same distance from a given centroid must have the same α values. However, real problems are often characterized by asymmetric clusters, and the asymmetries cannot be removed by any simple linear or nonlinear transformations of the data. The distance to the nearest neighbor then gives a more useful indicator of the degree to which a data point is anomalous.

Example 4.6 Consider applying a partitioning algorithm (such as k-means clustering) to the data set $\mathscr{D}_2 = \{-4, 1, 2, 3, 4, 5, 6, 7, 8, 9, 11, 14, 17, 20, 23, 26, 40, 41, 42, 43, 45\}$, with $k = 2$, yielding two clusters

$$C_1 = \{-4, 1, 2, 3, 4, 5, 6, 7, 8, 9, 11, 14, 17, 20, 23, 26\}$$

and

$$C_2 = \{40, 41, 42, 43, 45\}.$$

C_1 is not symmetric, unlike C_2. Since $\mu_1 \approx 10$, application of the distance-to-centroid criterion yields $\alpha(26) = 16$ whereas $\alpha(-4) = 14$, so that 26 appears to be the more anomalous of the data points. If we were to use the distance to the nearest neighbor as the criterion, on the other hand, we would infer that 26 is only three units away from its nearest neighbor, whereas -4 is 5 units away from its own nearest neighbor, hence the latter is more anomalous.

4.2.4 Boundary Distance

As pointed out earlier, nearest-neighbor based anomaly detection has its own pitfalls. In particular, if two outliers are near each other but far from all other data points, they need to be considered anomalous even though each is very close to the other. When anomaly detection relies on a preceding clustering step, the distance to the nearest cluster boundary provides another useful indicator of anomalousness, as illustrated in the following example.

Example 4.7 Consider the data set

$$\mathscr{D}_3 = \{-4, 1, 2, 3, 4, 5, 6, 7, 8, 9, 20, 21, 40, 41, 42, 43, 45\},$$

clustered using a non-partitioning algorithm with $k = 2$, yielding two clusters

$$C_1 = \{1, 2, 3, 4, 5, 6, 7, 8, 9\}$$

and

$$C_2 = \{40, 41, 42, 43, 45\}.$$

For the first outlier, $\alpha(-4) = 5$, the distance to the first cluster boundary, located at 1. On the other hand, $\alpha(20) = 11$ and $\alpha(21) = 12$, their respective distances from 9, the nearest cluster boundary. Hence 20 would be considered most anomalous, followed by 21, and then by -4, even though the nearest neighbor of 20 is only one step away.

If $\alpha(p)$ is to be assigned to points inside as well as outside a cluster, then this approach needs some modification. Recall that points inside a cluster may be least anomalous if they are distant from the boundary. One possible approach is to assign $\alpha(p) = 0$ for every point p that lies inside a cluster, and equals the distance to the nearest cluster boundary for points that lie outside clusters. But this assignment treats all inside points equally. However, if it is useful or necessary to evaluate the *relative anomalousness* of points within a cluster, we may instead define $\alpha(p) = -$(distance to the nearest cluster boundary) for points inside a cluster. This assignment is illustrated in the following example.

Example 4.8 As before, consider the data set

$$\mathscr{D} = \{-4, 1, 2, 3, 4, 5, 6, 7, 8, 9, 20, 21, 40, 41, 42, 43, 45\},$$

clustered using a non-partitioning algorithm with $k = 2$, yielding two clusters

$$C_1 = \{1, 2, 3, 4, 5, 6, 7, 8, 9\}$$

and

$$C_2 = \{40, 41, 42, 43, 45\}.$$

Note points -4, 20, and 21 do not belong to these two clusters. For points outside the clusters we have $\alpha(-4) = 5$, $\alpha(20) = 11$ and $\alpha(21) = 12$, representing their respective distances from the nearest cluster boundary points. For points inside the clusters, we have (for example)

$$\alpha(1) = 0, \alpha(2) = -1, \alpha(8) = -1, \alpha(40) = 0, \alpha(41) = -1.$$

This definition preserves the idea that points in the inner core of a cluster are to be considered less anomalous than points in the outer layers of the same cluster— we may imagine each cluster to be an onion, distinguishing outer layers from inner layers, rather than a strawberry whose interior is homogeneous.[2]

4.2.5 When Cluster Sizes Differ

When clusters have different sizes, as in the previous example, the choice of α function defined above considers points in the innermost core of the smaller cluster to be more anomalous than points in the innermost core of the larger cluster. For example, in the above example $\alpha(5) = -4$ whereas $\alpha(42) = -2$ although both are centroids of their respective clusters. This bias may be removed by normalization based on some function of the cluster sizes. One possibility is to multiply α values for the smaller cluster by $\max_{p \in C_i} \alpha(p) / \max_{p \in C_j} \alpha(p)$. This would result in the computation of $\alpha(42) = (-2)(-4/-2) = -4$, and $\alpha(41) = (-1)(-4/-2) = -2$.

If we use such a normalization step for points inside clusters, we then have to decide whether to use a similar normalization step for points outside clusters.

Example 4.9 Consider the data set

$$\mathscr{D} = \{1, 2, 3, 4, 5, 6, 7, 8, 9, 20, 40, 41, 42, 43, 45, 55\},$$

clustered using a non-partitioning algorithm with $k = 2$, yielding two clusters

$$C_1 = \{1, 2, 3, 4, 5, 6, 7, 8, 9\}$$

and

$$C_2 = \{40, 41, 42, 43, 45\}.$$

If the normalization step is performed for points inside the clusters, we have $\alpha(5) = \alpha(42) = -4$, as before. Without the normalization, we would have $\alpha(20) = 11$, the distance from 9 to 20, and $\alpha(55) = 10$, the distance from 45 to 55, i.e., 20 would be considered more anomalous than 55. But if the same normalization (as for interior points) is performed for points outside the clusters, we would have $\alpha(55) = (10)(-4/-2) = 20$, i.e., 55 would be considered more anomalous than 20.

Although a different normalization step would result in other answers, the fundamental question remains: *Should we consider characteristics of the "local" data space near a point, when evaluating the relative anomalousness of points (that lie outside clusters)?* A human needs to judge whether normalization is called for,

---

[2]If needed, a simple linear transformation can be defined to force all values to be positive, e.g., by adding the maximum cluster size to every α value 4.

e.g., whether 55 is truly more anomalous than 20 in the above example. The answer is context-dependent; in many applications, the characteristics of the local region needs to be taken into account, even when the data is one-dimensional.

For example, the heights of professional NBA basketball players have a larger variance than the heights of professional horse-racing jockeys, for biological reasons. So a professional basketball player who is almost 7 feet tall may be considered far less anomalous than a horse-racing jockey who is 5'6" (i.e., 5.5 feet) tall. Hence it makes sense to consider the distributions of the heights of the two sets of people, before computing the relative anomalousness of two outliers.

Another similar example is the set of salaries of individuals in a large organization, which can usually be clustered based on individuals' positions within the organization, e.g., executives make more money than managers, who make more than their administrative assistants, etc. But the variance in executive salaries is much higher than the variance in the salaries of administrative assistants, so that the magnitude of the deviation from the mean may not be a good indicator of relative anomalousness: an assistant who makes $40,000 more than others in the cluster (of assistant salaries) would be far rarer than an executive who makes $80,000 more than others in a different cluster (consisting of executive salaries).

In both of the cases above, normalization based on local data space characteristics would be useful. Other problems, thankfully, may involve no such complexity, e.g., when we compare scores of students in an examination, a student with a score of 100/100 may be considered as anomalous from the other high-performing students as the relative anomalousness of a student scoring 0/100 among other low-performing students.

4.2.6   Distances from Multiple Points

Many minor problems in data analysis disappear if we choose sample sizes that are not too small. The same holds for anomaly detection: if we consider distances to multiple (k) nearest neighbors, multiple cluster centroids, or multiple cluster boundaries, and consider the averages (or medians) over a collection of values, the results obtained tend to be more robust and less easily swayed by "noise" in the process that generated the data. The key question is then the choice of *how many* values need to be considered.

From the consideration of computational effort, of course, it is best to choose a small set of values over which averages (or medians) are computed. This leads to the common pragmatic choice of $k = 3$ (or 5) used by many practitioners, who reason that the additional advantage gained by using larger values of k is not worth the additional computational effort.

In problems where the computational effort considerations are not critical, however, the choice of k requires greater attention.

4.3 Conclusion

A large number of anomaly detection algorithms are based on clustering the given data, discussed in this chapter; such algorithms declare anomalies to be those data points that are outside clusters, or are near the boundaries of clusters. Although many practitioners restrict themselves to one or two popular algorithms, such as k-means clustering, there are many applications where the other algorithms discussed here are to be preferred, e.g., when the clusters in a dataset are not symmetric, or if densities vary across different regions of data space.

Although many clustering algorithms have been sketched in this chapter, we recognize that clustering is an extensively researched area, and problem-specific algorithms are sometimes developed. In particular, we sometimes must consider categorical (or nominal) data in which substantial changes are required to the "distance" measures on which algorithms such as k-means rely. An important set of clustering-like algorithms have been developed in the neural networks literature, e.g., the self organizing map (SOM) [77].

We conclude by remarking that the examples in this chapter illustrate the ideas using relatively simple and one-dimensional data. The problems discussed with such simple data are only exacerbated when we consider multi-dimensional data. Caution is recommended in developing complex solutions to specific problems without considerations for the generalizability of such approaches.

Chapter 5
Model-Based Anomaly Detection Approaches

Many data sets are described by *models* that may capture the underlying processes that lead to generation of data, describing a presumed functional or relational relationship between relevant variables. Such models permit comprehension and concise description of the data sets, facilitating identification of data points that are not consistent with such a description. This chapter explores anomaly detection algorithms based on hypothesizing and developing models that describe available data. The primary questions to be answered include the following:

- How is the model represented?
- How do we evaluate the extent to which a data point conforms to the model?
- What determines the occurrence of an anomaly?
- How do we derive or learn the model parameters from available data?

In Sect. 5.1, we discuss models describing the relationships between variables or attributes of data points, considering anomaly detection approaches based on variations in model parameter values, as well as variations from model predictions. In Sect. 5.2, we consider distribution models of data sets. Models of time-varying phenomena, and associated anomaly detection problems, are addressed in Sects. 5.3 and 5.4. This is followed in Sect. 5.5 by a discussion of the use of various learning algorithms to obtain models from data.

5.1 Models of Relationships Between Variables

When a dataset $\mathcal{D}$ is described by a model $M(\mathcal{D})$, anomaly detection can be performed either in the model parameter space or the data space. In the former case, the focus is on learned model parameters and how much they are influenced by a single data point. In the latter case, the relative anomalousness of different data points is measured by comparing the magnitudes of the variation between those

© Springer International Publishing AG 2017   57
K.G. Mehrotra et al., *Anomaly Detection Principles and Algorithms*, Terrorism,
Security, and Computation, https://doi.org/10.1007/978-3-319-67526-8_5

data points and model predictions. Section 5.1.1 discusses anomaly detection based on the effect of a data point on model parameter values. Section 5.1.2.1 addresses implicit and explicit relationships between model variables, assuming the data space approach.

5.1.1 Model Parameter Space Approach

In this approach, we evaluate how much the model is affected by the inclusion or exclusion of any given point in the data set, by comparing parameter values $\Theta = \{\theta_1, \ldots, \theta_k\}$ of the model of the entire dataset $M(\mathscr{D})$ with those of $M(\mathscr{D}_x)$, for some $\mathscr{D}_x \subset \mathscr{D}$ that depends on x. Multiple model parameter variations may be combined by defining the anomalousness of x to be

$$\alpha(x) = \sum_{i=1}^{k} |\Delta\theta_i|,$$

where $\Delta\theta_i$ is the difference between the parameter values of $M(\mathscr{D})$ and $M(\mathscr{D}_x)$. $\mathscr{D}_x$ may be defined by either excluding x from $\mathscr{D}$ or by considering a small subset of $\mathscr{D}$ that includes x. In the former (simpler) case, we may define $\mathscr{D}_x = (\mathscr{D} \setminus \{x\})$, which excludes the suspected outlier. But if the learning algorithm is robust enough to ignore outliers, or if the dataset is very large, then this approach may not work since we find that $M(\mathscr{D} \setminus \{x\})$ is no different from $M(\mathscr{D})$; then a better alternative would be to let $\mathscr{D}_x$ include x while being small enough for the model to be significantly affected by x, e.g., by restricting the size $|\mathscr{D}_x|$ to be a small fraction of $\mathscr{D}$, perhaps by random sampling and then explicitly including x.

The model parameter space approach requires that more than one model be learned from the data, which is more computationally expensive than approaches in which a single model is learned and variations from model predictions are measured in the data space. Nevertheless, the use of the model parameter space approach is preferable in some problems wherein the overall characteristics of the dataset are more important.

Example 5.1 Consider the set of annual incomes of individuals in the region of Seattle. An approximately normal distribution model may describe this set. But the exclusion of a few individuals (such as Bill Gates) may result in substantial change to the mean and standard deviation parameters of the normal distribution. Measuring the difference in mean (with vs. without any individual $x \in \mathscr{D}$) would convey the relative anomalousness of x. On the other hand, many high income individuals may not appear to be anomalous due to the effect of other outliers: a single billionaire's income significantly increases the mean and standard deviation.

In the above example, the model parameter space is more appropriate than a data space approach, since we are more interested in comparing x to $\mathscr{D} \setminus \{x\}$ rather than in modeling the relationship between attributes of x. The following example illustrates the application of this approach to a problem in which a model is described using a quadratic relationship between variables, along with an inequality constraint.

Example 5.2 Consider a problem modeled by quadratic function of the form $x_1^2 + x_2^2 + c = 0$ along with the inequality $x_1 x_2 + d > 0$, wherein variations in parameter values are combined linearly with equal weights in evaluating the relative anomaly score of any point, i.e., the anomaly score of a point is defined as $|\Delta c| + |\Delta d|$. Let the dataset $\mathscr{D}$ consist of ten points that are best described by values of parameters $c = -4.2, d = -0.4$, when a learning algorithm is applied to minimize mean squared error. Suppose $\mathscr{D}$ includes the points (1,2) and (2,0). When the same learning algorithm is applied to the dataset $\mathscr{D} \setminus \{(1, 2)\}$, let the values of the parameters be $c = -4.5, d = -0.4$, with the net difference in parameter values (compared to the original model) being $\alpha(1, 2) = |(-4.2 + 4.5)| + |(-0.4 + 0.4)| = 0.3$. On the other hand, when the same learning algorithm is applied to the dataset $\mathscr{D} \setminus \{(2, 0)\}$, let the values of the parameters be $c = -4.1, d = -0.1$, with the net difference in parameter values (compared to the original model) being $\alpha(2, 0) = |(-4.2 + 4.1)| + |(-0.4 + 0.1)| = 0.4$. Since this value is larger than 0.3, the point (2,0) would be considered more anomalous than (1,2).

5.1.2 Data Space Approach

The data space approach is conceptually simpler and less computationally expensive than the parameter space approach, as illustrated by the following example.

Example 5.3 Consider the set of two-dimensional data points

$$\mathscr{D} = \{(1, 1), (2, 1.9), (3, 3.5), (4, 4), (5, 5)\}.$$

Let a linear model $x_1 - x_2 = 0$ be constructed for $\mathscr{D}$. Three of the five data points have zero error with respect to this model, but $(2, 1.9)$ produces an error of 0.1, and $(3, 3.5)$ shows up with a slightly larger error (0.5), and is hence considered more anomalous.

In the above example, the data space approach is more appropriate since the best linear model constructed for $\mathscr{D}$ may be the same with or without the anomalous data points, depending on the methodology used to construct the model.

5.1.2.1 Implicit Model

We now address the data space approach for *implicit* models in which it is not possible to tease out explicit functional relationships. This method can be applied to problems that do not permit neat separation between the variables, e.g., it may not be possible to distinguish dependent and independent variables. But the data may be describable by implicit models such as $f(x) = \mathbf{0}$, possibly along with a set of constraints C consisting of inequalities $g_i(x) > 0$.

In such cases, the first possible measure of anomalousness of a data point x_0 is obtained by evaluating the distance between $f(x_0)$ and $\mathbf{0}$. We must then consider the extent to which the constraints are violated, and combine the two.

Example 5.4 Consider a dataset modeled by the equation $x_1^2 + x_2^2 - 4 = 0$ along with the inequality $x_1 x_2 - 0.5 > 0$. We evaluate the relative anomalousness of four points: (1.9, 2.1), (1, 2), (0.5, 0.6), and (2,0), using the data space approach.

- The first point almost exactly satisfies the first equation, with a small error, and easily satisfies the inequality. So it would not be considered anomalous, for any reasonable definition of anomalousness.
- The magnitude of the variation between (1, 2) from the first equation is $1^2 + 2^2 - 4 = 1$, whereas the inequality $1 \times 2 - 0.5 > 0$ does not contribute to the anomalousness measure.
- On the other hand, the magnitude of the difference between a point (0.5, 0.6) from the first equation is $|(0.5)^2 + (0.6)^2 - 4| = 3.39$, whereas $0.5 \times 0.6 = 0.3$ varies by $(0.5-0.3) = 0.2$ from the boundary described by the inequality. Since the violations from the equality as well as inequality are worse, this point would be considered more anomalous than (1,2).
- The point (2,0), on the other hand, produces zero error with respect to the first equation, while varying from the boundary described by the inequality by $|2 \times 0 - 0.5| = 0.5$, i.e., more than the similar variation of the previous point (0.5, 0.6).

The last two cases of this example raise the question of which point is more anomalous: (2, 0) or (0.5, 0.6)? The answer depends on the relative importance of the equality and the inequality describing the model, which may in turn be significantly problem-dependent. If a subject matter expert opines that variations from the inequality are twice as serious as variations from the equality, for instance, we would compute the composite anomalousness measure to be $3.39 + 2 \times 0.2 = 3.79$ for the point (0.5, 0.6), which is greater than $0 + 2 \times 0.5 = 1.0$ for the point (2,0). On the other hand, if the inequality variations are considered to be 20 times more important than the variations from the equality, then (2,0) would be considered more anomalous than (0.5, 0.6), since $0 + 20 \times 0.5 = 10 > 3.39 + 20 \times 0.2 = 7.39$.

In some cases, no problem-specific information is available a priori to decide how best to combine the multiple objectives represented by variations from the multiple equations and inequalities that describe the model. We must then infer the relative weights (of the variations from different constituents of the model) by examination of the data alone, possibly based on the high level statistical properties of these variations. Under the assumption of normal distributions for each of these variations, for instance, we may compute a data-dependent relative anomalousness measure as follows:

$$\alpha(x) = \sum_i \max(0, (d_i(x) - \mu_i)/\sigma_i)$$

where $d_i(x)$ is a non-negative quantity that measures the degree to which the ith model component (an equality or inequality) is violated by point x, μ_i is the mean of d_i over the entire data set, and σ_i is the standard deviation of the values of d_i. To avoid the distortions in μ_i and σ_i by strongly anomalous data points, we may instead use the median or the "trimmed" mean and standard deviation, e.g., omitting extreme values before calculating each average and standard deviation. The threshold for considering a value to be extreme may be arbitrary (e.g., the top 10% of data values before computing the mean), or based on quartile ranges, e.g., $x_{75\%} + 1.5 \times (x_{75\%} - x_{25\%})$, where $x_{p\%}$ refers to the smallest value that exceeds $p\%$ of the data set.

Many anomaly detection problems are focused only on abnormally high values of attributes or features, so that extremely low values are ignored. For example, if cash withdrawal magnitudes from bank accounts are being examined, only large withdrawals may come under scrutiny. However, there are some applications in which low values are also important, e.g., if the goal is to find instances of low productivity in the workplace.

5.1.2.2 Explicit Models

In explicit models, the model identifies the relationships between certain *dependent* variables y and other *independent* variables x. A data point $(x_1, \ldots, x_m, y_1, \ldots y_n)$ that does not closely conform to the model is to be considered anomalous, while most data points are expected to be largely consistent with the model.

The relationship between dependent and independent variables is presumed to be expressible in the closed form $y = F(\Theta, x)$, where F represents the model for the dataset $\mathcal{D}$, Θ refers to model parameters that are learned for dataset $\mathcal{D}$, and $(x, y) \in \mathcal{D}$ refers to a specific data point. The definition of an anomalousness measure is then straightforward, based on evaluating the variation between predicted and actual data values:

$$\alpha(x, y) = |y - F(\Theta, x)|.$$

Well-known distance measures can be used to evaluate the extent to which a specific data point conforms to a model. For example, if $(x_1, x_2, x_3, y_1, y_2)$ is a given data point, and (f_1, f_2) is the model intended to explain the last two variables as functions of the first three variables, then the values of the following expressions are possible choices to evaluate how closely the point $p = (x_1, x_2, x_3, y_1, y_2)$ conforms to the explicit model (f_1, f_2):

- Manhattan: $\alpha_1(p) = |y_1 - f_1(x_1, x_2, x_3)| + |y_2 - f_2(x_1, x_2, x_3)|$
- Euclidean: $\alpha_2(p) = \left((y_1 - f_1(x_1, x_2, x_3))^2 + (y_2 - f_2(x_1, x_2, x_3))^2\right)^{\frac{1}{2}}$

- Generalized: $\alpha_3(\boldsymbol{p}) = w_1(|y_1 - f_1(x_1, x_2, x_3)|)^a + w_2(|y_2 - f_2(x_1, x_2, x_3)|)^a$ where $a > 0$ and the weights w_1 and w_2 express the relative importance of the two dependent variables, or normalize the summands so that they have magnitudes in the same ranges.

If we use the above expressions of $\alpha_i(\boldsymbol{p})$ directly to measure the anomalousness or outlierness of $\boldsymbol{p}$, we would be assuming that (f_1, f_2) is an accurate model of the given data. However, these models may themselves be imprecise or uncertain, so that small variations from the model should not be considered significant. For instance, a satisfactory linear relationship between dependent and independent variables does not preclude the possibility that none of the data points has zero error with respect to the linear model. In this case a small error can be ignored as explained below. If the degree of imprecision of the model can be estimated from the data, e.g., if it is known that 90% of the data points vary from a model by $\alpha(\boldsymbol{p}) < \epsilon$, then we may define the anomalousness of a point $\boldsymbol{p}$ using this known (and permitted) range of variation ϵ, e.g., using an expression such as the following:

$$\alpha_{\beta,\epsilon}(\boldsymbol{p}) = (\max(0, \alpha(\boldsymbol{p}) - \epsilon))^{\beta}$$

where $\beta > 1$ indicates the rate at which distance exacerbates anomalousness.

A special case of explicit models arises in the context of classification problems, with the dependent variable y taking values that indicate class membership, where the number of classes is usually small. The models are expected to be simple with a small number of parameters (e.g., perceptrons that separate classes using a linear function of the input variables), and data points misclassified by the model can be considered to be anomalous, with the degree of relative anomalousness measured in terms of the shortest distance to the appropriate boundaries between the classes. Euclidean norms are usually chosen for the distance (or error) measure, since simplicity and differentiability permit gradient computations and the application of gradient descent algorithm variants to minimize such distances.

Example 5.5 Consider the set of points $\{(1,1), (2,1), (3,0), (4,1), (5,1), (6,1), (7,0),$ $(8,0), (9,0), (10,0), (11,0), (12,0), (13,1), (14,0), (15,0)\}$, where the second variable is intended to be a function of the first, and is a class membership indicator, i.e., the points in $\{(1,1), (2,1), (4,1), (5,1), (6,1), (13,1)\}$ are associated with class 1, and the others with class 0.

If we restrict ourselves to models of the form "$y = 0$ iff $x > c$" with a single parameter c and error measure $\sum_i |(c - x_i)|$, the best model for this dataset is obtained with the parameter choice $c = 6$ or any value of $c \in [6.0, 7.0)$. In other words, $c \in [6.0, 7.0)$ minimizes an error measure such as $\sum_i |(c - x_i)|$ for the samples misclassified by any choice of c. Of the two misclassified data points $(3,0)$ and $(13,1)$, the former is at a distance of 3 from $c = 6$ and is hence considered less anomalous than the latter which is at a distance of 7. However, if we measure anomalousness using a different error measure such as $\sum_i (c - x_i)^2$ (for misclassified samples), then note that $c = 6.9$ is better than 6.0, since $(3 - 6.9)^2 + (13 - 6.9)^2 < (3 - 6)^2 + (13 - 6)^2$. Even more surprisingly, optimizing this error measure gives a better solution at $c = 7.9$, which misclassifies one additional point (7,0). In this example, minimizing mean squared error gives a different result from minimizing the number of misclassified data points.

5.2   Distribution Models

In this section, we first discuss the application of parametric distribution estimation to anomaly detection, and then address regression models of multiple kinds.

5.2.1   Parametric Distribution Estimation

Instead of focusing on the relationships between individual attributes of data points, we may address compact descriptions of entire sets of data, viewed as distributions with some expected variations among each other. Although this section focuses on the well-known univariate normal distribution, our discussion can be applied to various other distributions studied in the literature. If the data is believed to follow a Gaussian distribution, then the probability density is given by the expression

$$f(x; \mu, \sigma^2) = \frac{1}{\sqrt{2\pi\sigma^2}} \left(e^{-\frac{(x-\mu)^2}{2\sigma^2}} \right).$$

The distribution has a characteristic peak that occurs at the mean, μ. Then the main task is to learn the most appropriate values of μ and σ (standard deviation), the two parameters of this distribution. About 4.5% of the data points are expected to be more than two standard deviations away from the mean, and about 0.3% are expected to be more than three standard deviations away. 99.9% of the data points are expected to lie within the range $\{\mu - 3.29\sigma, \mu + 3.29\sigma\}$, while 99.99% of the data points are expected to lie within the range $\{\mu - 3.89\sigma, \mu + 3.89\sigma\}$; hence $|x - \mu|/\sigma$ gives a clear indication of the probability of expecting the occurrence of a point. Great caution is required in interpreting these numbers when the distribution is not normal, or when the dataset is too small to justify the assertion that distribution is normal; not every unimodal (bell-shaped) distribution is normal.

   One of the primary tasks is then to distinguish between the expected amount of *noise* in the data from the occurrence of true outliers. For instance, if a dataset is represented by a normal distribution, then almost every point is expected to deviate from the mean, but this does not indicate that each such point is anomalous. Indeed, if the distribution is normal, we expect that a certain fraction of the points may lie far from the mean, although such fractions are expected to become very small as we move farther away from the mean. The tricky question is whether we would consider a point situated at $> \mu + 3\sigma$ to be anomalous (for a one-dimensional data set), given that the very definition of a normal distribution suggests that there is a non-zero probability of finding a few such points.[1]

The following are a few alternative approaches to this conundrum:

1. Ignore the definition of the distribution, and use the relative distance from the mean to indicate relative anomalousness of each data point with respect to the mean.
2. Based on the definition and properties of the distribution, use the probability of occurrence of a specific data point to indicate relative (non)-anomalousness of points.
3. Compare the distribution parameters with and without the suspected anomalous data point, using the relative difference between such parameters to indicate relative anomalousness.

For anomaly detection applications in which some data points cluster together, and a small number do not, *mixture models* are most appropriate. A mixture model consists of a finite collection of parameterized distributions, e.g., Gaussian, binomial, exponential, or log-normal distributions.

Example 5.6 House prices in a city vary widely, but the city consists of a collection of neighborhoods, with considerable uniformity within each neighborhood. The overall distribution of house prices is hence a combination of the distributions within its various neighborhoods; each such distribution may be normal, with a different mean and variance.

Parametric models assume that the available data can be described by a well-understood formula or distribution (e.g., Gaussian), with a small number of parameters whose values need to be learnt for the specific data under consideration. If this assumption is justified, then efficient learning algorithms can be applied to the available data; otherwise, we have to rely on less efficient non-parametric approaches.

5.2.2 Regression Models

Most regression algorithms apply variations of a gradient descent algorithm in which an error measure is successively reduced in magnitude through iterative improvements, i.e., changing the coefficient values in the model in a direction that would reduce error. This approach, focusing on the least mean square approach, was first developed by Legendre (in 1805) and Gauss (in 1809), who applied it to infer planetary orbits.

---

[1]In the multivariate case, deviations must be normalized by standard deviations of multiple attributes.

5.2.2.1 Linear Regression

Linear regression is a special case which assumes that each dependent variable is a linear function of the independent variables; variants in the neural network literature are referred to as the *Perceptron* and *Adaline* models. They usually focus on finding the linear model parameters that minimize mean squared error, i.e., $MSE = \sum_i (y_i - f(x_i))^2$, where y_i is the dependent variable, x_i is the ith independent variable (vector), and $f(x_i) = a_0 + \sum a_j x_{i,j}$ is the linear function model. The values of a_j parameters are easily obtained, *e.g.,* by iteratively modifying each linear coefficient a_j by $-\eta d(MSE)/da_j$, where $\eta > 0$ is a small *learning rate* (or *step size*) constant chosen to be small enough to avoid divergence, but large enough to ensure progress. Linear models may also be learnt by minimizing another error measure such as $\sum_i |(y_i - f(x_i))|$, although gradient descent cannot be applied easily.

5.2.2.2 Nonlinear Regression

Linear models are limited in their capabilities, although easy to formulate and learn. Methods such as *non-linear least squares* extend this approach to nonlinear functions, such as $f(x_i, \beta)$ where β represents the parameter vector. Often, the primary difficulty is that closed form solutions are not available. Nevertheless, the values are successively improved over multiple iterations, e.g., using a Taylor series expansion. Initial parameter values may be obtained by trial and error, and iterations are terminated when progress over successive iterations is judged to be negligibly small.

If **e** represents the error vector, and J is the Jacobian matrix consisting of first partial derivatives of the error vector with respect to the parameter β, then in each iteration β is updated by $(J^T J)^{-1} J^T \mathbf{e}$.[2] When the Jacobian itself cannot be computed easily, numerical approximations are obtained by computing the effects of small perturbations of the parameter β.

Another important approach for nonlinear regression is the use of *Feedforward Neural Networks* [85], whose parameters (referred to as *weights*) are learned using the *error backpropagation* algorithm (abbreviated *backprop*) [117]. Pictorially, they can be represented as in Fig. 5.1, which shows an input layer (representing the independent variables), and output layer (representing the dependent variables), and a *hidden layer*; the input and hidden layers are augmented by a dummy node x_0 whose constant output is 1, whose outgoing edge weights are also referred to as bias values. Each edge from node i in layer $k - 1$ to node j in layer k is annotated by

---

[2]For the computation of this quantity, the Gauss-Newton method relies on the Cholesky decomposition method, expressing any positive-definite matrix as the product of a lower triangular matrix and its conjugate transpose. The Gauss-Newton method is susceptible to divergence, a problem addressed by reducing the magnitude of the change in β, e.g., by using the *Marquardt parameter*. The Levenburg-Marquardt algorithm adapts the magnitude of this update, effectively obtaining a compromise between the Gauss-Newton algorithm and gradient descent.

Fig. 5.1 Feedforward Neural
Network with a hidden layer

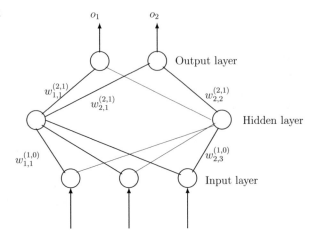

a numeric *weight* parameter $w_{j,i}^{(k)}$, and each node in the network applies a sigmoid function (such as the hyperbolic tangent or the logistic function) to the net input received by the node. The output of the jth node in the kth layer, for instance, is

$$x_j^{(k)} = 1/(1 + e^{-(\sum_i w_{j,i}^{(k)} x_i^{(k-1)})}).$$

Such a network with a sufficient number of nodes can be used to approximate any desired smooth function to a required degree of accuracy. A stochastic gradient descent procedure (backprop) iteratively updates weights, starting with the layers closest to the outputs and propagating backwards until the layers near the input nodes are updated. Theoretical results state that two hidden layers are sufficient for approximation to any desired degree of accuracy, with each node connected to each node in the next layer, although recent work has exploited the power of "deep networks" with many layers and a small number of connections between successive layers.

5.2.2.3   Kernel Regression and Support Vector Machines

This approach estimates model functions using a locally weighted average function, where the weights are determined by a suitably chosen *kernel* function[91, 114]. Gaussian functions are often used as kernels, an approach that has been popularized with Support Vector Machines (SVMs) [29, 105, 112] which have also been used for classification problems. SVMs have emerged as the learning algorithms of choice for many classification problems, and combine several theoretically important ideas (such as regularization, nonlinear feature space representations, and interior point optimization algorithms) captured in efficient implementations. An SVM with Gaussian kernel functions resembles a neural network model known as a *Radial*

Basis Function (RBF) network, whose nodes apply Gaussian functions to net inputs, and the parameters of such functions are learnt by gradient descent algorithm variants.

5.2.2.4   Splines

Formally, a spline is a piecewise-polynomial real function; i.e., the function $S(t)$ consists of a collection of polynomials $P_i(t)$, with smooth junctions where the pieces connect. A frequent choice of the polynomial function $P_i(t)$ is a cubic polynomial. Splines patch together multiple polynomial functions, where each polynomial's coefficients are determined by a few points in close proximity to each other. In the context of time series modeling, discussed in detail in the next section, time is broken up into k discrete intervals $[t_0, t_1), [t_1, t_2), \ldots, [t_{k-1}, t_k]$, and the spline $S(t) = P_i(t)$ for the range $t_{i-1} \leq t < t_i$, where $1 \leq i \leq k$.

5.3   Models of Time-Varying Processes

Formally, a *(univariate) time series* is a totally ordered sequence of data items (numerical values), each associated with a time-stamp which makes it possible to identify the time gap between any two items. Two successive timestamps are not required to have the same time gap, e.g., the time interval between two different pairs of successive items in the same time series may be 10 s and 20 s, respectively, since we cannot always assume that data is sampled at equal time intervals. The format of the time-stamp does not matter, as long as there is a one-to-one correspondence with a universally accepted clock. In a *multivariate* time series, data items are vectors of numerical values.

It may appear to be possible to consider the time-stamp of a data item as just another numerical attribute of data, and hence apply the anomaly detection principles and algorithms used with non-time series data. However, this approach is unlikely to succeed since time is a unique characteristic that cannot be treated like other attributes of a data set. Timestamps serve to sequence the data items (vectors of non-time attributes of the data) as well as to indicate the "spacing" between data items along the temporal dimension.

Example 5.7   Consider the following univariate time series data represented as two-dimensional vectors (x_i, t_i) in which the second attribute is a time-stamp:

$$\mathcal{D}_t = \{(-1, 0), (1, 1), (3, 2), (5, 3), (7, 4), (6, 5), (11, 6), (17, 9)\}$$

- If we were to attempt to find an anomaly while ignoring the fact that the second attribute represents a time-stamp, the last point (17,9) is at the greatest

Euclidean distance from other points, and is hence considered to be anomalous. This conclusion would emerge irrespective of which of the anomaly detection approaches (discussed in other chapters) we apply.

- If we were to treat the first attribute as a function of the second, the data fall neatly into a straight line pattern, described by the equation $x = 2t - 1$, except for one outlier: (6,5). In the 2-dimensional data space, this point appears to be squarely in the middle of a cluster, although it is substantially anomalous from the time series perspective.

The above example illustrates the importance of separating out the time attribute. It is not necessary that the other data attributes be functions of time, and generally they are not. For example, the daily fluctuations in closing stock prices cannot be described as functions of time, affected as they are by multiple factors of the economy and company performance.

The time series in the above example had data items that were not regularly spaced along the time dimension. Note that if regular spacing did occur in the time dimension, e.g., if the dataset had also included (13,7) and (15,8), the apparent anomalousness of (17,9) would be eliminated, but the time dimension would then be non-informative with respect to non-time-series anomaly detection approaches; (6,5) would still appear to be in the heart of the data set, hence not anomalous using such approaches.

Hence it is necessary to develop anomaly detection algorithms that are specifically applicable to time series problems. A natural starting point is to extend or modify non-time-series approaches, as explored below.

A time series, $\mathscr{X}$, can be expressed as an ordered sequence:

$$\mathscr{X} = (x_1, t_1), (x_2, t_2), \ldots, (x_n, t_n) \text{ or}$$

$$\mathscr{X} = x_1, x_2, \ldots, x_n$$

In either case, x_i denotes the data attribute(s) at time t_i; and can take continuous or discrete values. The former representation is typically employed when t_i values are not equally spaced, e.g., when sensor measurements are reported at varying intervals. In the rest of the chapter we assume that the data points are equally spaced and use the compact representation, i.e., $\mathscr{X} = x_1, x_2, \ldots, x_n$.

Several methods have been proposed in recent years to address model-based anomaly detection. Model-based methods, such as Regression[43, 101], Auto-Regression [45], ARMA [94], ARIMA [87], and Support Vector Regression [82] may be used to find abnormal sequences. Chandola et al. [21] suggest a kNN based method which assigns anomaly score, equal to Euclidean distance to kth nearest neighbor in the data set, to each time series. Another distance measure based approach that can be used with kNN is dynamic time warping (DTW) proposed by Berndt and Clifford [11]. Fujimaki et al. [45] suggest Autoregressive (AR) approach which constructs a global AR model for all series and then calculates the anomaly score at time t as the gap between the observed value and the predicted value. Zhou et al. [124] propose to apply the Back propagation Neural Network to the results of Local Outlier Factor (LOF), to analyze the outliers over data streams.

Most of the anomaly detection techniques for sequence data, discussed in this chapter, can be grouped into three categories, [21]): kernel-based techniques, window-based techniques and Markovian techniques.

- **Kernel-based techniques** compute a similarity measure based on the entire time series or sequence, which may be used in conjunction with the anomaly detection techniques developed for non-sequence data, such as the k-nearest-neighbor approach.
- **Window-based techniques** analyze short windows of sequence data (subsequences) to compute an anomaly score for each subsequence, then combine these to obtain a total anomaly score.
- **Markovian techniques** assign a probability to each subsequence based on previous observations. For a system whose prior states were $S_1, \ldots, S_t$, Markovian approaches assume that the conditional probability of the next state depends only on $k < t$ recent states, i.e.,

$$P(S_{t+1}|S_1, \ldots, S_t) = P(S_{t+1}|S_{t-k+1}, S_{t-k+2}, \ldots, S_t).$$

For the simplest Markov models, $k = 1$, i.e., only the immediately preceding system state determines the probability of the system state.

With each of these models, anomaly detection consists of identifying significant variations from a model's predictions, relying on a similarity or distance measure that either compares, as stated earlier:

(a) predicted data to actual data, or
(b) the parameters of models with vs. without the potentially anomalous data point.

First a model is generated to predict the behavior of the time series; using this model, the predicted values are calculated and compared with the observed values. The cumulative score of the differences is defined as the anomaly score of each observed data object. This approach is sketched in Algorithm "Model-based approach":

Algorithm Model-based approach

1: GIVEN: Time series dataset $\mathscr{X} \in \mathscr{D}$, parametric model M;
2: Train the model M on each time series $\mathscr{X}$ in $\mathscr{D}$ to determine appropriate parameter values, and let the result be M;
3: **for** each $\mathscr{X} \in \mathscr{D}$ **do**
4:    Apply M to predict values of $x(t)$ of time series $\mathscr{X}$ and evaluate distance associated with it;
5:    Report the series $\mathscr{X}$ as an anomaly if its distance is substantially large;
6: **end for**

Stationarity assumptions are sometimes made, i.e., assuming that the process which generates the data has not changed with time.

5.3.1   Markov Models

Markov models are used in systems that pass through sequences of discrete states (such as $s_1 s_2 s_3 s_2 s_4 s_1 s_2 \ldots$). Markov models presume that each individual's future state has some dependence on the current state, but not directly on earlier states. Time and other variables, including possible states, are assumed to be discrete, e.g., with values for the time variable ranging over non-negative integers, and values for the possible states of an individual or system ranging over a finite set S.

Notation Let S_t denote the state of the system at time t, and let $p_{t+1,b}$ denote the probability that the system state will be b at time $t+1$. Three simple Markov models are presented in Fig. 5.2. In the simplest Markov model, $p_{t+1,b}$ is a function of S_t.

Example 5.8 Consider a coin being tossed repeatedly, with $S_t \in \{Heads, Tails\}$ indicating that a coin toss at time t results in one of two alternatives.

- With a fair coin, we expect $p_{t,Heads} = p_{t,Tails} = 0.5$, easily observed using a coin tossed repeatedly; this is a trivial case of a Markov chain. But if multiple coin tosses reveal a different probability distribution, e.g., if Heads show up 70% of the time, we suspect an unusual behavior compared to the behavior of most coins which are expected to be unbiased or fair. If the number of coin tosses is small, the anomaly may be attributed to randomness, but if the anomaly was observed over a large number of tosses (say 100), then we suspect that the coin toss process is biased (not fair).
- The coin toss may be anomalous in other ways, even if almost the same number of heads and tails are reported. Runs of identical coin-toss results may occur, e.g., with $p_{t+1,Heads} = 0.8$ if $S_t = Heads$, and $p_{t+1,Tails} = 0.8$ if $S_t = Tails$. Discovery of this pattern may be obtained by applying a learning algorithm that learns the probability of the result of each successive coin toss as a function of the most recent coin toss.

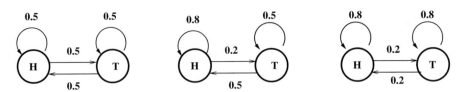

Fig. 5.2 Three HMM models with conditional probability of H and T as indicated by *round arrows*; a *straight arrow* provides the probability of transition from one state to another. The leftmost figure is for a fair coin, the second is biased (in favor of Heads), and the third figure is symmetric but has low probability of going from one state to another

- After the discovery of this pattern, a player may temporarily substitute a fair coin, whose behavior (with equiprobable outcomes of every coin toss) would constitute an anomaly with respect to the learned model. But we would not consider this to be an anomaly if this substitution is permanent, i.e., if the fair coin continues to be used thereafter.
- The "bias" introduced into the coin may wear off with time, and observations after a few hundred coin tosses may vary from the 0.8 model, with the magnitude of such variation from the model measured over multiple coin tosses. If this effect (reduced bias) persists, it would reflect a necessary change in the model, but not an anomaly.

Summarizing, this approach would first construct a Markov model for the system being observed, and anomalies would be identified by the occurrences of temporary variations from expected behavior in an observed system. This could be measured in two ways:

1. Anomaly detection may be based on evaluating the magnitude of the variation between the observed behavior and the behavior of the system as predicted by the Markov model.

Example 5.9 In a repeated coin toss example, let the values observed be in the sequence HHHHT HTHHH HHHTHH TTHHH HHHHH THHHH HHHTHH HTHTT TTHHH TTHHT HTTHT HHTHH TTHHH HHHHH THHHH. Then the Markov model would predict that Heads are likely to occur with a probability of 0.7, as shown in Fig. 5.3. The stretch of equiprobable outcomes in the center of this sequence would be flagged as anomaly since it significantly varies from the 0.7 prediction; this can be seen by the dip in the center of the graph in Fig. 5.4.

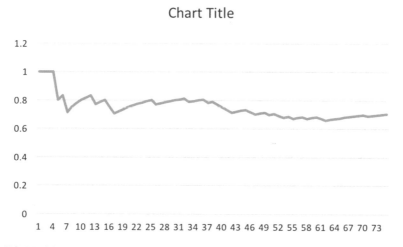

Fig. 5.3 Model parameter *Prob(Heads)* based on all preceding time points

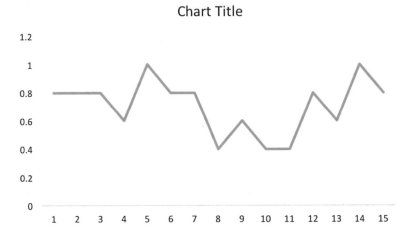

Fig. 5.4 Result of periodically updating the parameter *Prob*(*Heads*) after 5 coin tosses

2. Instead, anomaly detection may occur upon constructing a new model of the
   system (e.g., by continuously learning the parameters of the Markov model), and
   then computing the differences in model parameters from the previously learned
   parameters. Since learning is computationally expensive, the model parameters
   may be recomputed periodically instead of being continuously updated.

Example 5.10 In the preceding coin toss example, if the model parameter
that describes the probability of the next coin toss resulting in a Heads
outcome, updated at a frequency of a hundred coin tosses, has the values
$(0.80, 0.79, 0.81, 0.79, 0.79, \mathbf{0.52}, 0.80, 0.81, 0.79, \ldots)$, the an anomaly would be
flagged at the occurrence of the 0.52 parameter value.

5.3.2   Time Series Models

Time series are sequences of data items carrying time tags that monotonically
increase, usually at a fixed rate, e.g., stock prices at closing time on each work
day.[3]
    Past work has extensively addressed the identification of autoregressive rela-
tionships, (long-term) trends, and cycles. Smoothing techniques such as moving
averages have been used to suppress the effects of random noise in the data. All of
these characteristics are derived from the available data, and expressed as parameters
of a high level model such as *ARIMA*, discussed in detail in textbooks such as [54].

---

[3]Note that the same phrase "time series" may refer to a single time-ordered sequence of values or
a set of such sequences.

Sometimes, the behaviors of individuals and systems need to be analyzed at the high level, rather than applying a microscope to irrelevant fluctuations over time at a low level of granularity. Well-known statistical approaches to analyze time series can then be useful, e.g., describing each time series (over a given period of time) in terms of cycles and long-term trends. These can be viewed as the primary features of the time series, and used to evaluate the occurrence of an anomaly within a single time series as well as to distinguish one time series from a collection of others.

- A retail company's stock may have previously seen seasonal cycles, e.g., with an annual high in December. If, in one December, a comparable high is not encountered, this signals that something anomalous is occurring within that company.
- A collection of oil company stocks may exhibit a strong upward trend over a relatively long period, despite occasional dips or corrections. If one of these stocks fails to show such a trend, it stands out as anomalous with respect to others.
- The rate at which accesses to a database occur may follow patterns that are repeated daily, e.g., peaking early morning and early afternoon, with prominent dips around noon and in the late afternoon. Variations from this, e.g., a very late afternoon peak, may indicate an anomaly such as a user inappropriately accessing an account.

Example 5.11 Consider the sales charts for a retailer, illustrated in Fig. 5.5.

Fig. 5.5 Example monthly sales chart for a retail company

- First, we observe that there is a steady upward trend in the sales, despite occasional minor fluctuations attributable to various external causes or the random vagaries of human wills in numerous individual purchase decisions. A linear regression model shows an annual increase of about 2%; if sales in a three-month period show flat sales (0%) in a year-to-year comparison, this indicates an anomaly with respect to the linear trend model for prior data, even if a similar phenomenon has occurred occasionally in the past.
- Second, there are some seasonal patterns in monthly sales, with December sales being approximately twice November sales each year. If sales in a particular year show little or no increase in December, compared to the prior month, this would indicate an anomaly with respect to the seasonality model.

Many real life processes are better described using continuous-valued variables which change their values along a linear time sequence. It is possible to discretize continuous-valued variables (e.g., into *high, medium, low* discrete categories), but this often involves loss of information, and also requires problem-specific knowledge. For example, an investor is interested in knowing actual stock prices, and it is not sufficient to say merely that a stock price is "high." Time series have hence been used extensively in many applications in dynamic system modeling and forecasting, e.g., in financial markets, and have been studied for many years, [15].

The main effort in statistical approaches is to find the right model that describes a time series; anomalies within the time series are then considered to be data points that deviate substantially from the predictions of such a model.

5.3.2.1 ARIMA Models

ARIMA models are very widely used, and allow us to capture the following features of many time sequences:

- Dependence of a variable upon its own history;
- Greater weightage to its recent values than older previous values;
- Non-stationarity;
- Smoothing by moving averages, to eliminate some noise;
- A term that expresses drift over time; and
- Inclusion of terms that represent random noise.

Dependence on past values is a common assumption in time series modeling. To model dependence, the current value of a variable, x_t, is described in terms of its past values; $x_{t-1}, x_{t-2}, \ldots$. An *autoregressive* AR (p) model assumes that we can predict the behavior of x_t as a linear function of its past values $x_{t-1}, x_{t-2}, \ldots, x_{t-p}$, where $p > 0$ and the associated AR(p) model is

$$x_t = \sum_{i=1}^{p} \phi_i x_{t-i} + \epsilon_t,$$

where the *error* ϵ_t is assumed to be normally distributed with mean 0 and variance σ^2. The *lag* operator L is often used to refer to the value of a variable at a preceding time point, with $Lx_t = x_{t-1}$; for longer time lags, we write $L^i x_t = x_{t-i}$. Using this convenient notation, the above AR(p) equation can instead be written as follows:

$$\left(1 - \sum_{i=1}^{p} \phi_i L^i\right) x_t = \epsilon_t.$$

Often the current level is affected by a *shock* variable which is randomly distributed, whose effect is felt for an interval of time stretching over $q > 0$ time units. This is an example of a phenomenon captured effectively by using a *Moving Average (MA)* model. For example, a MA(1) model is written as

$$x_t = \epsilon_t + \theta\epsilon_{t-1}$$

and an MA(q) model is:

$$x_t = \epsilon_t + \theta_1\epsilon_{t-1} + \ldots + \theta_q\epsilon_{t-q} = \left(1 + \sum_{i=1}^{q} \theta_i L^i\right) \epsilon_t.$$

When the AR(p) and MA(q) models are combined, we get a general ARMA(p, q) model:

$$\left(1 - \sum_{i=1}^{p} \phi_i L^i\right) x_t = \left(1 + \sum_{i=1}^{q} \theta_i L^i\right) \epsilon_t.$$

Finally, introducing a *drift* term with an additional parameter $d > 0$ that allows us to capture non-stationarity in the time sequences, we obtain the *Autoregressive integrated moving average* or *ARIMA(p,d,q)* models, proposed by Box and Jenkins, summarized in [8], and represented by the following equation:

$$\left(1 - \sum_{i=1}^{p} \phi_i L^i\right) (1 - L)^d X_t = \delta + \left(1 + \sum_{i=1}^{q} \theta_i L^i\right) \epsilon_t$$

where p, d, q are model parameters that may be chosen by the user. Here, $\delta/(1 - \sum_i \phi_i)$ is considered the "drift" over time. Special case of this model includes Random walk model in which $p = q = 0$. A special case, *ARFIMA(0,d,0)*, written as $(1 - L)^d X_t = \epsilon_t$, is interpreted to mean

$$\epsilon_t = X_t - X_{t-1}d + X_{t-2}d(d - 1)/2 - X_{t-3}d(d - 1)(d - 2)/6 + \ldots$$

Important variations of ARIMA are: VARIMA models, in which the data varying over time are vectors; SARIMA models that explicitly model seasonality; and

FARIMA or ARFIMA models, in which parameter d can take fractional values, so that deviations from the mean can be allowed to decay more slowly.

5.3.2.2   Discrete Fourier Transformation

The discrete Fourier transform (DFT) represents a collection of digitized signals (time series) in frequency domain (sinusoids). Fourier transforms are important in signal processing. due to the reason that it allows to view the signals (discrete time series) in a different domain where several difficult problems become very simple to analyze.

Given a time series $\mathcal{X} = x_0, x_1, \ldots, x_{n-1}$, the discrete Fourier transformation of $\mathcal{X}$ is a n-dimensional vector, $\mathcal{F} = F(0), F(1), \ldots, F(n-1)$, defined as follows:

$$F(j) = \sum_{k=0}^{n-1} x_k e^{-i\frac{2\pi}{n}jk}; j = 0, 1, \ldots (n-1),$$

where i denotes the (complex) square root of -1. The transformation can be written in a matrix notation as:

$$\mathcal{F} = W \mathcal{X}^T$$

For example, when $n = 4$ and $\mathcal{X} = (10, 3, 7, 5)$, in matrix notation the desired transformation is:

$$\begin{bmatrix} F(0) \\ F(1) \\ F(2) \\ F(3) \end{bmatrix} = \begin{bmatrix} 1 & 1 & 1 & 1 \\ 1 & -i & -1 & i \\ 1 & -1 & 1 & -1 \\ 1 & i & -1 & -i \end{bmatrix} \begin{bmatrix} 10 \\ 3 \\ 7 \\ 5 \end{bmatrix} = \begin{bmatrix} 25 \\ 3 + 2i \\ 9 \\ 3 - 2i \end{bmatrix}.$$

Given the Fourier transformation $\mathcal{F}$ of a given time series it is very easy to recover the series by inverse transformation.

5.3.2.3   Haar Wavelet Transformation

We now describe Discrete Wavelet Transformation (DWT) with Haar wavelets (also known as Db1). Haar wavelets are the simplest possible wavelets, preferred over DFT due to computational efficiency and the time localization property, so that it is easier to determine when the anomaly occurs (Fig. 5.6)

Given a vector of x of 2^n points, where n is an integer, its Haar transformation is:

$$y = Hx,$$

Fig. 5.6 Basis functions of
Haar transformation, for data
with four observations

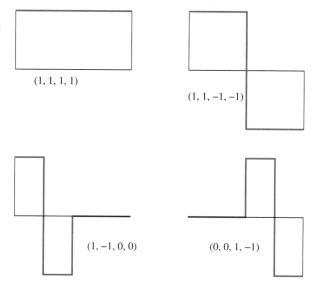

where H is an orthogonal matrix with special properties, described below. For
example, if x has four points, then the graphical representation of the basis functions
of this transformation are as shown below:This basis can be represented by four
vectors: $(1, 1, 1, 1)$, $(1, 1, -1, -1)$, $(1, -1, 0, 0)$ and $(0, 0, 1, -1)$. These vectors are
orthogonal to each other, however they are not orthonormal. The matrix H is a 4×4
matrix, obtained after normalization of these vectors. More precisely,

$$H_4 = \frac{1}{2} \begin{bmatrix} 1 & 1 & 1 & 1 \\ 1 & 1 & -1 & -1 \\ \sqrt{2} & -\sqrt{2} & 0 & 0 \\ 0 & 0 & \sqrt{2} & -\sqrt{2} \end{bmatrix}$$

For example, when $n = 4$ and $\mathscr{X} = (10, 3, 7, 5)$, in matrix notation the desired
Haar transformation is:

$$\frac{1}{2} \begin{bmatrix} 1 & 1 & 1 & 1 \\ 1 & 1 & -1 & -1 \\ \sqrt{2} & -\sqrt{2} & 0 & 0 \\ 0 & 0 & \sqrt{2} & -\sqrt{2} \end{bmatrix} \begin{bmatrix} 10 \\ 3 \\ 7 \\ 5 \end{bmatrix} = \begin{bmatrix} 12.5 \\ 0.5 \\ 7\frac{1}{\sqrt{2}} \\ \sqrt{2} \end{bmatrix}$$

Similarly, when x contains eight observations, the corresponding basis functions
are as plotted below (Fig. 5.7). The corresponding transformation matrix, H_8 is given
below:

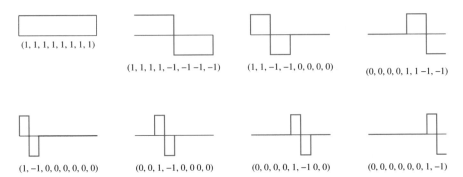

Fig. 5.7 Basis functions of Haar transformation, for data with eight observations

$$
H_8 = \frac{1}{\sqrt{2}}
\begin{bmatrix}
1 & 1 & 1 & 1 & 1 & 1 & 1 & 1 \\
1 & 1 & 1 & 1 & -1 & -1 & -1 & -1 \\
\sqrt{2} & \sqrt{2} & -\sqrt{2} & -\sqrt{2} & 0 & 0 & 0 & 0 \\
0 & 0 & 0 & 0 & \sqrt{2} & \sqrt{2} & -\sqrt{2} & -\sqrt{2} \\
2 & -2 & 0 & 0 & 0 & 0 & 0 & 0 \\
0 & 0 & 2 & -2 & 0 & 0 & 0 & 0 \\
0 & 0 & 0 & 0 & 2 & -2 & 0 & 0 \\
0 & 0 & 0 & 0 & 0 & 0 & 2 & -2
\end{bmatrix}
$$

In general, unnormalized transformation matrix H_{2n} is recursively defined as:

$$
H_{2n} = \begin{bmatrix} H_n \otimes [1, 1] \\ I_n \otimes [1, -1] \end{bmatrix},
$$

where I_n is the identity matrix of size $n \times n$ and $\otimes$ denotes the Kronecker's product.

Haar transformation has the property that if $y = Hx$, then $x = H^{-1}y = H^T y$. Consequently, if we retain all values in y, then x can be recovered completely. However, if only larger values in y are retained and all small values are equated to zero, then the inverse transformation provides a 'good' approximation of x.

More recently, other methods to find a model for a given time series have also been proposed, such as short-time Fourier transform and fractional Fourier transform.

5.4   Anomaly Detection in Time Series

Anomaly detection problems come in two primary flavors in the context of time series:

- Abnormalities within a single time series; and
- Substantial variation of one time series from a collection of time series.

This section is organized as follows. In Sect. 5.4.1 we consider anomaly detection within a single time series; and the variation of one time series from a collection of time series is discussed in Sect. 5.4.2.

5.4.1 Anomaly Within a Single Time Series

The problem of anomaly detection within a single time series is usually formulated as the "Change Detection" problem, i.e., identifying when the parameters of a time series model have changed.

We first present some examples of time series anomaly detection problems (Figs. 5.8, 5.9, 5.10).

Example 5.12 The daily closing price of a large retail company's stock has been exhibiting a roughly linear behavior for many months, with small up and down variations.

- One day, the stock price rises a little, but may fall the next day back to the value predicted by the linear behavior. This may not be an anomaly, and may be attributed to market noise, especially if the magnitude of the rise is not substantial compared to the usual day-to-day variations for the same stock price.
- One week, a sudden spike is observed, not followed by a dip over the next few days. This indicates an anomaly, and could be explainable by news suggesting that the fundamentals of the company's performance have substantially improved.
- In early December, the stock price rises substantially, and then falls again by early January. This may not be an anomaly, but may indicate cyclic seasonal behavior (due to Christmas-time spending), especially if supported by data for multiple years. Of course, we may consider any such December behavior to be anomalous with respect to the behavior in other months. Thus, the time duration

Fig. 5.8 Daily closing prices for a company's stock in 2015

Fig. 5.9 Daily closing stock prices, with an anomaly near day 61

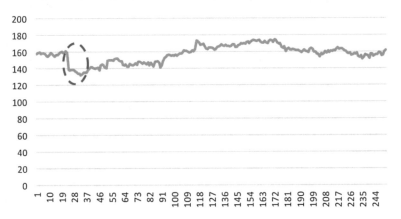

Fig. 5.10 Daily closing stock prices, with an anomaly that extends for several days (roughly from day 22)

of the analysis is important to distinguish seasonal or cyclical variations from true anomalies.

- On September 16, 2008, the stock price dipped substantially. This certainly indicates an anomaly in the time series. Later analysis reveals that practically all stock prices dipped on that day (referred to as a "market crash"), and this particular stock had merely moved in lock-step with the rest of the stock market. Hence, when viewed in the context of the rest of the stocks, the behavior of this stock was not anomalous.

- On May 6, 2010, the stock price does not change much, with respect to its earlier observed linear behavior. But analysis reveals that most large company stock prices dipped on that day, and this particular stock had not moved in lock-step with the rest of the stock market. Hence, when viewed in the context of the rest of the large company stocks, the behavior of this stock was anomalous!

In the context of cyber-security, time series arise in a few different contexts:

- The observable behaviors of an individual may exhibit changes over time, which can indicate that he now has new intentions that he did not have earlier, leading us to infer the possibility of past or future fraudulent activities.

Example 5.13 Alberta Einstein's normal access pattern for her email account may be characterized by the fact that she checks in for only a few minutes each night at 9PM. If it appears that Alberta's account is being used for over fifteen minutes at midnight, we may infer that perhaps the account is being accessed by an unauthorized user, not Alberta. Of course, it is possible that there are valid reasons for such occurrences, e.g., travel to a different time zone. If this recurs with regularity every Tuesday night (but not on other nights), perhaps Alberta's computer has a zombie problem.

- The behavior of an individual may vary substantially from those in his peer group. This "outlier" individual may be involved in fraudulent activities.

Example 5.14 Johann Dough, an employee of a hospital with legitimate access to electronic healthcare records may be accessing the healthcare records of seventy two individuals in an eight-hour day during a certain extended duration of time. On average, other individuals with the same job title at the same hospital access forty individuals' records per day, with a standard deviation of ten. Johann's substantial variation from peer group behavior signals possible undesirable activity, e.g., Johann may be leaking confidential information to external entities.

- If the job description of Johann was different (e.g., "network administration" or "system testing") from the peer group, then the variation may not be considered significant.
- If the standard deviation was twenty instead of ten, the "three-sigma" rule suggests that the same variation may not be considered to be substantial, since $(70 - 40)/20 < 3$.
- Suppose Johann had been previously accessing fifty records per day, with a sudden jump to seventy two at some point in time; when that jump occurs, the variation with Johann's own past history signals an anomaly necessitating further investigation. Of course, this does not conclusively prove fraudulent behavior, e.g., such behavior is explainable if Johann's job description had changed recently, or if new software (improving productivity) had been purchased at the time point when the transition in Johann's behavior occurred.
- Suppose Johann was steadily accessing exactly 8–9 records in every hour of the weekday, whereas the peer group exhibits more predictable fluctuations, e.g., peaking at 9AM, with a near-zero access rate during the lunch hour, and a slow period towards the end of the work-day. This would signal an abnormality when you compare the hourly access rate time series of Johann with the time series of the peers. (A possible explanation is that an automated process or zombie is performing the accesses.)
- Suppose Johann's access rate steadily drops over a long period of time, whereas peer group access rates remain flat. This is again an indication of

an anomaly of Johann's time series vs. the peer group's time series, though it may indicate loss of morale, motivation, or productivity, rather than fraud. Abnormally low levels of activity hence require different actions (and the formulation of different underlying hypotheses) when compared to anomalies indicating significantly higher levels of activity.

- The quantities of message traffic received over the Internet by a large server or system may exhibit patterns that are mostly predictable, although subject to some random noise. Substantial variations from such patterns (over time) may signal possible cyber-attacks.

Example 5.15 The typical distribution of the number of email messages received by a server for an e-retail sales company (providing online purchase assistance) may peak to 100 messages/second at 9PM, with very low activity over most of the day, and a spike during the noon hour.

  – If a peak is instead observed at 2AM on a specific day, that would be an anomaly with respect to the past history of the message traffic.
  – If the peak volume at 9PM jumps to 1000 messages/second, that would indicate a different kind of anomaly, e.g., the announcement of a "hot" new toy on the retail channel, or the precursor of a denial-of-service attack.
  – The peak on a Saturday may be at a different time, with weekend hourly time series following a substantially different pattern than weekday time series. So an anomaly for a weekday may not be an anomaly for weekends (or holidays).
  – Data from a long period of time may be necessary to distinguish anomalies from normal business cycle events, e.g., due to year-end accounting activities or major holidays associated with extensive gift-related purchases or specially discounted sales.

- The distribution of the origins or routing points of Internet message traffic may also follow predictable patterns, with malicious attacks identifiable by variations from such patterns.

Example 5.16 Approximately 3% of the message traffic to a university server usually originates from a specific country, and has a predictable distribution over time, e.g., with 80% of the traffic originating during the period 9AM–6PM in that country. A potential cyber-attack may be suspected when the actual distribution of traffic from that country deviates drastically from this pattern.

  – One morning, 10% of the message traffic received by the university originates from that country. The substantial increase in volume indicates possibility of malicious attacks or other undesirable activity.
  – Another day, 0.01% of the message traffic originates from that country. This anomaly may indicate that the communication channels specific to that country are broken.
  – A non-trivial amount of traffic from that country is received at a time period corresponding to midnight in that country. The amount may not be substantial when viewed in isolation, but may deviate from the expected time distribution.

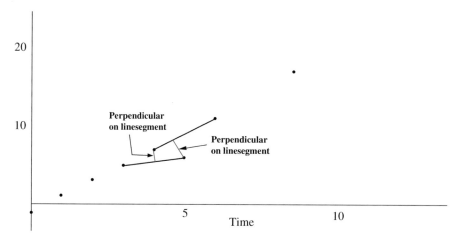

Fig. 5.11 Distance of data point (12, 12) using one consecutive point on either side

– One day, the distribution of lengths of messages (from that country) appears
  to be unusual, e.g., many messages have the same length, whereas the
  norm is to see much greater variation in message length. This suggests that
  many messages may be coming from the same sender, indicating potential
  spamming or malicious activity.

5.4.1.1 Methodologies for Anomaly Detection Within a Single Time Series

Many non-time-series problems are amenable to distance-based approaches, where
the relative anomalousness of a data point is estimated as a function of its Euclidean
distance to other data points (the nearest neighbor or the nearest cluster centroid),
which is not directly useful for time series data due to the complications introduced
by time-stamp values. This leads to the alternative distance-based approaches listed
below.

Notation The terms "preceding" (or "predecessor") and "succeeding" (or "successor") refer to the nearest points on either side of a point in the given data set,
along the time dimension; for instance, in the time series the data point (11,6) is
preceded by (6,5) and followed by (17,9). We use the phrase "data item" to refer
to the attributes in a data point excluding the timestamp, e.g., "6" is the data item
corresponding to the data point (6,5) where "5" is the timestamp.

• Compute the perpendicular distance between a data item and the line segment
  connecting its predecessor and successor data item.[4]

---

[4]The distance between (x_1, t_1) and the line segment connecting its predecessor (x_0, t_0) and
successor (x_2, t_2) can be computed as

This is undefined for the first and last time points.

Example 5.17 For $\mathscr{X} = \{(-1,0),\ (1,1),\ (3,2),\ (5,3),\ (7,4),\ (6,5),\ (11,6),\ (17,9)\}$, the respective distances for the non-terminal points ($t = 1,\ldots,6$) are approximately 0.0, 0.0, 0.0, 1.3, 1.3, and 0.7, respectively, leading to the conclusion that the most anomalous points are (7,4) and (6,5) as shown in Fig. 5.11.

Note that the presence of an anomalous data point such as (6,5) affects its neighborhood, leading us to suspect that its nearest neighbor is also anomalous. This is a consequence of using only two points to construct the line segment, for which the distance of a potentially anomalous point is computed.

- Compute the distance between a data points and the line segment connecting its two nearest (in time) data points. This permits computation of the relative anomalousness of even the terminal points, although requiring extrapolation instead of interpolation in some cases. Implicitly this permits the possibility that the time series behaves differently in different regions of the temporal space, e.g., with a different slope.

Example 5.18 For $\mathscr{X}' = \{(-1,0),\ (1,1),\ (3,2),\ (5,3),\ (12,12),\ (17,15),\ (17,16),\ (17,17)\}$, the respective distances for all points are approximately 0.0, 0.0, 0.0, 0.0, 5.0, 0.0, 0.0, and 0.0, respectively, leading to the conclusion that the most anomalous point is (12,12).

In the above example, the first part of the time series has a positive slope, whereas the second part has zero slope; the anomalous data point was in between these segments. The extreme points within each segment were not considered anomalous since they did lie close to (or on) the line segments constructed from their nearest neighbors.

- We may use $k > 2$ nearby points to construct the line segments, e.g., the best fit line connecting two previous and two following data points, and then compute the distance as above; see Fig. 5.12.
- The distance of a point could be computed from a curve that is used to describe the time series, where the curve is obtained by regression or *splines*.

5.4.2   Anomaly Detection Among Multiple Time Series

When two time series can be represented satisfactorily by ARIMA(p, d, q) models, and the series-defining parameters ($\tilde{\phi}, \tilde{\theta}$) of the models are estimated using available time series data, one notion of distance (between the two time series) can be obtained in terms of the differences in the series-defining parameters ($\tilde{\phi}, \tilde{\theta}$), e.g.,

$$\frac{\|(t_2 - t_0)x_1 - (x_2 - x_0)t_1 - x_0t_2 + x_2t_0\|}{\sqrt{(x_2 - x_0)^2 + (t_2 - t_0)^2}}.$$

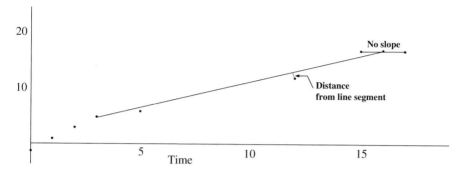

Fig. 5.12 Distance of data point (12,12) from the regression line obtained using two consecutive neighbors on either side of it

the Euclidean distance between the $(\tilde{\phi}, \tilde{\theta})$ vectors for the two time series models. However, there are a few disadvantages; the numbers of the AR and MA parameters must be equal, and in addition, even a small difference in p, d, q parameters of two ARIMA models can result in substantially different time series.

The same principle can be applied to other models, e.g., when two time series are represented using cubic splines with different coefficients, we can compute the distance between the vectors of coefficients for the two time series.

It is a more complex task to compute the distance between a single time series $\mathcal{X}$ and multiple time series (over the same dataset $\mathcal{D}$). If the parameter vectors of the models of all the time series in a set $\mathcal{D}$ are substantially similar, and all are described by the same ARIMA class (with the same p, d, q values), then averaging the $(\tilde{\phi}, \tilde{\theta})$ vectors of multiple time series may be sufficient to capture the characteristics of the entire collection of time series, so that we can compute the distance between the parameters of $\mathcal{X}$ and the mean (vector) of the parameter vectors of all the time series in $\mathcal{D}$.

On the other hand, if there are substantial differences among the models of the time series in $\mathcal{D}$, e.g., a few of them have slightly different p, d, q values, then this approach will not work. Instead, the original dataset $\mathcal{D}$ would have to be averaged out, yielding a single time series $\mathcal{X}_{\mathcal{D}}$, for which an ARIMA$(p, d, q)$ model is derived, and the resulting $(\tilde{\phi}, \tilde{\theta})$ vector compared with the vector for the best ARIMA(p, d, q) model for $\mathcal{X}$; if the distance is large, we conclude that $\mathcal{X}$ is anomalous with respect to $\mathcal{D}$. This procedure requires fixing p, d, q values based on the best ARIMA model obtained for the entire data set.

A similar approach may also be used to identify an anomaly within the same time series $\mathcal{X}$ that ranges over time from t_1 to t_n, such as when the behavior of the underlying process (generating the time series data) changes.

Notation Let $\Theta(\mathcal{X})$ represent the $(\tilde{\phi}, \tilde{\theta})$ vector corresponding to the best ARIMA(p, d, q) model for the time series $\mathcal{X}$, with the best possible choice of p, d, q. For the same p, d, q values, let $\Theta_{j,k}(\mathcal{X})$ represent the $(\tilde{\phi}, \tilde{\theta})$ vector corresponding to the ARIMA(p, d, q) model for the time series $\mathscr{X}$ when restricted to the sub-series from time t_j to time t_k.

A few alternative ways of estimating the anomalousness of $\mathscr{X}$ at time t_j are as follows:

- Compute the distance between $\Theta(\mathscr{X})$ and $\Theta_{j,\|\mathscr{X}\|}(\mathscr{X})$.
- Compute the distance between $\Theta(\mathscr{X})$ and $\Theta_{j,j+\ell}(\mathscr{X})$ for a specified sufficiently long interval ℓ.
- Compute the distance between $\Theta_{j-\ell-1,j-1}(\mathscr{X})$ and $\Theta_{j,j+\ell}(\mathscr{X})$ for a specified ℓ.

We return to the question of determining *how close is a time series to a collection of other time series*. For the rest of this section, we assume that $\mathscr{D}$ is the collection of m time series, $\mathscr{X}_i$ is the ith time series in $\mathscr{D}$, for $1 \leq i \leq m$, and the goal is to define an appropriate distance measure $d(\mathscr{X}_i, \mathscr{D})$, to be maximized by any anomaly detection procedure.

Before we can define such a measure, a few issues need to be addressed:

- Two time series may be defined over overlapping but not identical periods of time. In this case, the distance measure may focus only on the period for which both time series are defined, i.e., the intersections of the two time periods.
- One or more of the time series may be irregularly spaced, with missing values. If no other information is available, as discussed in an earlier section, missing values may be filled in by linear interpolation between data points on both sides of the missing values, with the hope that the rate of change is constant during the missing period. Interpolation using a collection of neighboring points may be advisable, since anomalies may distort the results of interpolation for neighboring points.
- Two time series may range over different sets of data values, e.g., one company's stock value may vary around \$10 while the other varies around \$50. Normalizing all time series to the same range of values, e.g., the interval [0,1], would facilitate comparison. The simplest normalization approach is to replace each point $x_{\ell,i} \in \mathscr{X}_\ell$ by a time series-specific normalization, such as

$$\frac{(x_{\ell,i} - \min_j[x_{\ell,j}])}{(\max_j[x_{\ell,j}] - \min_j[x_{\ell,j}])}$$

which ensures that all the time series have the same [0,1] range.[5] Alternatively, we may focus on making all time series have the same mean (0) and standard deviation (1); this is accomplished by replacing each point $x_{\ell,i} \in \mathscr{X}_\ell$ by

---

[5]This normalization can be excessively influenced by a single very large (or very small) value, which can be addressed by *censorization* or trimming. A simple censored (or trimmed) normalization would remove points that are not in the 10-percentile to 90-percentile range within that time series, before applying the normalization transformation mentioned above. As a result, most normalized values would be in the [0,1] range, while a few may vary substantially from this interval.

$$\frac{x_{\ell,i} - \frac{1}{|\mathscr{X}_\ell|} \sum_j x_{\ell,j}}{\sigma(\mathscr{X}_\ell)}$$

where σ refers to the standard deviation within a given time series.

In the following, for presentational and notational convenience, we assume that all $x_{\ell,i}$ are normalized; all time series are of length n, and $t = 1, 2, \ldots, n$. Finally, the dataset $\mathscr{D}$ contains m different time series.

5.4.2.1 Using Point-to-Point Distances

The first approach is to define the distance of a time series from others, as the average of distances of points in the time series from other time series. This can be obtained by first calculating point by point distance between two time series, averaged over all times; i.e., defined as:

$$d(\mathscr{X}_i, \mathscr{D}) = \frac{1}{n} \sum_{t=1}^{n} \sum_{j=1}^{m} d(x_{i,t}, x_{j,t})$$

We assume here that all the time series are being compared over the same time period, so that $|\mathscr{X}_i| = |\mathscr{X}_j| = n$.

5.4.2.2 Using Variations over Time

A time series is more than a collection of points: it conveys increases and decreases that occur over time, which may be more significant than the values themselves. So the right question to ask may be whether the increases or decreases in two time series correlate strongly. The "local gradient" of a time series at a given point in time can be estimated, and compared with the local gradients of other time series at the same point in time. Two time series that differ substantially in their data values may vary over time in the same manner and hence have high similarity.

Example 5.19 Let $\mathscr{X}_1 = \{(1, 1), (4, 2), (6, 3), (3, 4), (1, 5)\}$ and $\mathscr{X}_2 = \{(7, 1), (8, 2), (10, 3), (5, 4), (4, 5)\}$. Both of these time series "rise" at the first two time points and then "fall" at the next two time points, and are hence substantially similar even though the actual data points are substantially different. On the other hand, $\mathscr{X}_3 = \{(1, 1), (4, 2), (2, 3), (3, 4), (1, 5)\}$ exhibits a decrease at the second time point and an increase at the third time point, although it is almost identical to $\mathscr{X}_1$. Viewed as sets of points, $\mathscr{X}_1$ and $\mathscr{X}_3$ are almost identical; viewed in terms of local gradients, on the other hand, they are substantially different.

Depending on the nature of the problem and the granularity of the desired gradient analysis, the effort required in estimating the local gradient can be varied, as discussed below (Figs. 5.13, 5.14).

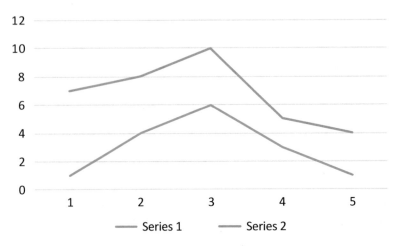

Fig. 5.13 Synchronous rise and fall of $\mathscr{X}_1$ and $\mathscr{X}_2$

- In the simplest case, we can ignore the magnitudes of changes, and only consider whether data is rising, falling, or flat. Define the local gradient of $\mathscr{X}$ at time t as $x_{t+1} - x_t = \Delta x_t$. Then the local or instantaneous correlation between two series $\mathscr{X}$ and $\mathscr{Y}$ at time t is defined as

$$\rho_s(\mathscr{X}, \mathscr{Y}, t) = \frac{1}{2}\left((sgn(\Delta x_t) + sgn(\Delta y_t))\right),$$

where the three-valued 'signum' function is defined as:

$$sgn(x) = \begin{cases} 1 & \text{if } x > 0 \\ -1 & \text{if } x < 0 \\ 0 & \text{if } x = 0 \end{cases}$$

The value of $\rho_s(\mathscr{X}, \mathscr{Y}, t)$ ranges over $\{-1, -0.5, 0, 0.5, 1.0\}$, and can be converted to a distance measure $\in [0, 1]$ as follows:

$$d_s(\mathscr{X}, \mathscr{Y}, t) = \frac{1}{2}\left(1 - \rho_s(\mathscr{X}, \mathscr{Y}, t)\right).$$

The distance between $\mathscr{X}$ and $\mathscr{Y}$ over the entire time period $(|\mathscr{X}_i| = |\mathscr{X}_j| = n)$ is

$$d_s(\mathscr{X}, \mathscr{Y}) = \sum_{t=1}^{n} d_s(\mathscr{X}, \mathscr{Y}, t).$$

- The above measure does not distinguish between the magnitudes of changes in the variables (over time). High resolution measurements of the data over time

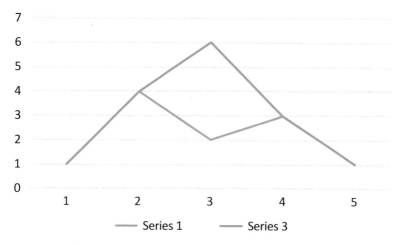

Fig. 5.14 Asynchronous behavior of $\mathscr{X}_1$ and $\mathscr{X}_3$

may result in rendering system noise to be visible, so that tiny fluctuations result in apparent up-and-down movement of the variable being studied. No correlation may be observable when the variables in both time series fluctuate in this manner, distorting the computation of the distance measure, although real changes of substantial magnitude occur in the variables over time. For example, the directions of minute-by-minute fluctuations in two stocks of the same kind may not expose the correlations that are visible in day-by-day stock price measurement variations. Thus, it is important to consider the magnitudes of changes, in addition to the directions. But the notions of "large" and "small" must be considered with respect to each specific time series; for instance, the magnitude of the daily variation in a volatile company stock may be much larger than that of most other stocks. The normalization approaches mentioned earlier normalize the magnitudes of data points, e.g., into the [0,1] range, but not the variations. To address these concerns, another measure is defined, normalizing by the average deviation magnitude, as follows:

$$\rho_c(\mathscr{X}, \mathscr{Y}, t) = \frac{1}{2}\left(\frac{\Delta x_{t+1}}{\mu_{dxt}} + \frac{\Delta y_{t+1}}{\mu_{dyt}}\right),$$

where $\mu_{dxt} = \sum_{t=1}^{n}|\Delta x_{t+1}/(n-1)|$ and $\mu_{dyt} = \sum_{t=1}^{n}|\Delta y_{t+1}/(n-1)|$.
As before, we can then define the distance measure

$$d_c(\mathscr{X}, \mathscr{Y}, t) = (1 - \rho_c(\mathscr{X}, \mathscr{Y}, t))$$

and

$$d_c(\mathcal{X}, \mathcal{Y}) = \sum_{t=1}^{n} d_c(\mathcal{X}, \mathcal{Y}, t).$$

A discretization approach may also be used, categorizing changes that occur in a time series into a few classes. Changes whose magnitudes are of the order of "system noise" may be considered to be zero, and others may be characterized as small, medium, or large, using problem-specific thresholds inferred from the data.

5.4.2.3   Correlations with Delays

In some practical problems, two time series may exhibit similar behavior but one time series may be lagging another, e.g., when variations in the inputs of a manufacturing plant lead to similar variations in the plant's outputs after the time lag required to complete processing the inputs.

Formally, let $\mathcal{Y}$ represent the time series $\mathcal{X}$ when shifted by Δt time units, i.e., with $y_t = x_{t+\Delta t}$. We seek to find the value of Δt that minimizes $d_c(\mathcal{X}, \mathcal{Y})$. The new *delayed-distance* measure is defined as

$$d_d(\mathcal{X}, \mathcal{Y}) = min_{\Delta t} d_c(\mathcal{X}, \mathcal{Y}).$$

This computation is expensive, and various heuristics may be used to obtain reliable estimates of the same.

One minor problem, especially with relatively short time series, is that the least delayed-distance may be obtained by shifting one time series almost all the way to the end of the other, e.g., $(4, 1), (2, 2), (3, 3), (4, 4)$ will have zero distance from $(1, 1), (2, 2), (3, 3), (4.4)$ after shifting by three time units, even though the unshifted time series are substantially similar. This can be addressed by penalizing the delayed-distance measure by the amount of shift required, or by accounting for the relative lengths of the sequences that match after the shift occurs.

These methods have been used mainly for detecting an individual outlier ($w = 1$), not for abnormal subsequence detection, and the results are impacted significantly by model parameters or distributions of data sets. How do we define anomaly score for a subsequence $(x(i), x(i + 1), \ldots, x(i + w - 1))$? An example, based on this approach, is as follows:

$$\alpha\left(x(i), x(i + 1), \ldots, x(i + w - 1)\right) = \sum_{j=i}^{i+w-1} \alpha(x(j))$$

where, for some $\epsilon > 0$,

$$\alpha(x(j)) = \begin{cases} 1 & \text{if } |x(j) - x^*(j)| > \epsilon \\ 0 & \text{otherwise.} \end{cases}$$

The subsequence $(x(i), x(i+1), \ldots, x(i+w-1))$ is considered to be anomalous if

$$\alpha \, (x(i), x(i+1), \ldots, x(i+w-1)) > w \times \theta,$$

where the value of the threshold $\theta \in [0.0, 1.0]$ indicates the relative preference for false positives *vs.* false negatives, e.g., $\theta = 0.1$ is expected to yield more false positives than $\theta = 0.3$.

When the window length $w = 1$, this approach is essentially that of checking whether the difference between a predicted value and the actual value exceeds a prespecified threshold. This approach [82] has been applied using a Support Vector Machine (SVM) [105] trained to predict the next value in the sequence from the immediately preceding subsequence of length τ.

A similar approach can be used for subsequence anomaly detection using the Discrete Fourier transform (DFT) or Discrete Wavelet transformation (DWT), as outlined below.

The Haar transform is obtained for subsequence $(x(i), x(i+1), \ldots, x(i+w-1))$ of size w for $i = 1, 2, \ldots, n-w+1$ and a predetermined number $(<< n)$ of wavelet coefficients are retained. Using these retained coefficients the subsequences that are considered to be sufficiently similar to a given sequence Q are obtained. The set S of neighbors of Q, consisting of those subsequences whose distances do not exceed a preassigned threshold, is obtained; thus facilitating detection of subsequences that are farther away.

Anomaly detection, using association rules and model based approach, is also possible [58]; in particular, when the *inputs* and *outputs* of a system can be distinguished and data mining methodologies can be used. In this approach, a finite number of representative patterns are first obtained from each continuous time series by random sampling of subsequences of fixed lengths and their features, such as minimum, maximum, and the first few DFT coefficients. Using these features, the subsequences are clustered, and cluster centroids are considered the *items* for association rule mining. A typical association rule, $X_1 \to Y_2$, is interpreted as saying that the presence of pattern X_1 in the time series X leads to the prediction that the pattern Y_2 is likely to appear in series Y. These association rules can be used to detect anomalies, if X_1 appears in X along with some subsequence in Y that substantially varies from pattern Y_2.

5.5 Learning Algorithms Used to Derive Models from Data

Models are sometimes available based on the prior knowledge of subject matter experts, but more often the models need to be inferred from available data by the application of learning algorithms. However, learning algorithms constitute an entire research area by themselves; this section is hence confined to only a small subset of the same.

We must first choose from among a class of possible models, while considering three criteria: simplicity, adequacy (in terms of minimal error), and computational effort. *Occam's razor* suggests that the simplest model be chosen, with the fewest possible parameters; this is often justified from the perspective of generalizability, *e.g.,* the ability of the model to describe new data points derived by the same procedure that generated the observable data points. Expert knowledge is first invoked to determine which class of models is appropriate for the task, and which learning algorithm may be most useful; in the absence of such knowledge, empirical experimentation is carried out with multiple models and learning algorithms, in order to find the simplest model that can explain the data with minimal error, using an acceptable amount of computational time for the learning algorithm. The criterion of minimizing error magnitude is often at odds with the other two criteria (model simplicity and learning effort), so compromises are desired, perhaps with thresholds of acceptability for each criterion. The practitioner must decide whether a chosen model is adequate enough to describe the data, and consider the trade-offs between these conflicting criteria.

Models are usually obtained by applying a learning algorithm to a subset of available data (referred to a "training data"), and then evaluated using another subset (the "test data").

We have considered three different kinds of models in the preceding sections, each of which calls for a different class of learning algorithms.

- **Distributions:** Learning algorithms that attempt to find models for data distributions (e.g., to determine if a normal distribution fits the data well) have been well studied in the statistical literature, with some estimates for the extent to which a dataset may deviate from the mathematical description of the model.
- **Time-varying phenomena:** Algorithms for the estimation of Markov models and ARIMA time series models for data have also been described in the statistical literature.
- **Relationships between variables:** Most learning approaches for describing the relationships between variables can be regarded as regression or curve-fitting algorithms, in which model parameters are to be estimated from training data.

5.5.1 Regularization

As mentioned earlier, simplicity, adequacy , and computational effort are three essential characteristics of all learning algorithms. Regularization plays an important role in accomplishing these goals, as explained below in context of mean squared error (regression).

In addition to minimizing the mean squared error, often an important concern is to ensure that coefficient values in the model remain small. This is often accomplished by penalizing large coefficients, for which a few different approaches have been explored [57, 111]. For example, in *Ridge regression*, we minimize a linear combination of the mean squared error and the sum of the square of the coefficients, instead of just the mean squared error. The effect is to shrink all coefficient values.

A related approach is *Lasso (Least Absolute Shrinkage and Selection Operator) regression* [110] which restricts the sum of the magnitudes of the coefficient values to remain below a predetermined threshold, with the net effect of forcing some coefficients to be zero-valued. Ridge regression is analogous to setting the prior distributions of coefficients to be normal, whereas Lasso regression uses prior distributions that are sharply peaked at zero values, with discontinuous first derivatives.

Elastic net [30] regularization combines the approaches of ridge and lasso regression, minimizing a function such as

$$|Y - AX|^2 + \lambda_1 \sum_i |A_i| + \lambda_2 \sum_i |A_i|^2$$

Other variations have also been proposed, e.g., penalizing large changes over successive iterations, or enforcing co-existence of non-zero values for some variables (known to be strongly correlated). In all of these methods, the choice of the regularization coefficients λ_i is critical to obtain fast computation without divergence.

Forward-backward (or *Proximal Gradient*) methods [27] have recently been proposed to address problems in which the penalty terms are non-differentiable. For Lasso regression, this approach leads to a simple update scheme for model parameters, in which potential parameter updates are first computed using the gradient approach $(w - \eta d(MSE)/dw)$, and the results are subjected to a *soft thresholding* step so that

$$w_{new} = \begin{cases} w - \eta d(MSE)/dw - \eta & \text{if } w - \eta d(MSE)/dw > \theta \\ w_{new} = 0 & \text{if } |w - \eta d(MSE)/dw| \leq \theta \\ w_{new} = w - \eta d(MSE)/dw + \eta & \text{if } w - \eta d(MSE)/dw < -\theta \end{cases}$$

where θ is a threshold. Iteration of this update rule converges well with an appropriate choice of η.

5.6 Conclusion

Mathematical modeling techniques have been developed for various problems, for understanding data, representing essential aspects of data, and generating predictions for as-yet unobserved data. Many kinds of models exist, and their applicability depends on the nature of the data and the kinds of problems to be addressed. Models can also be used for anomaly detection, in two ways: either focusing on the effect of a data point on the model parameter values, or on the variation between actual data and the model predictions.

This chapter has summarized the use of different kinds of models for anomaly detection. Models bring order to the universe of data: as the volume and complexity of data increases, models become more important since it can become very difficult to analyze the data without the models. A particularly important example is that of data that vary over time, where the relationships between prior and current values are critical in determining whether observed data is anomalous. We have discussed the possible use of well-known methodologies such as Markov models and ARIMA models for anomaly detection.

Model parameter learning methodologies have also been discussed in the preceding section. From our perspective, even hitherto unexplored learning algorithms can be used to build models, which can then be used for anomaly detection. Finally, we caution that expert knowledge may be useful in first determining what kind of model is appropriate for a given dataset, as well as for determining high level constraints on parameter values, and the relative importance of various features of the data, in order to avoid emphasizing spurious or accidental relationships..

Part II
Algorithms

Chapter 6
Distance and Density Based Approaches

In Chap. 3, we discussed distance based approaches for anomaly detection; however there the focus was to illustrate how distances can be measured and minor perturbations in proposed distance can change the outcome; illustrated by simple examples. In this chapter we consider anomaly detection techniques that depend on the distances and densities. The densities can be global or local to the region of concern.

6.1 Distance from the Rest of the Data

The simplest anomaly detection algorithms are based on the assumptions about the data distribution, e.g., that data is one-dimensional and normally distributed with a mean of $\bar{p}$ and standard deviation of σ. A large distance from the center of the distribution implies that the probability of observing such a data point is very small. Since there is only a low probability of observing such a point (drawn from that distribution), a data point at a large distance from the center is considered to be an anomaly. More specifically, in such simple cases, anomaly detection algorithms may rely on the fact that, as z increases, the number of data points found at a distance of $z\sigma$ away from $\bar{p}$ decreases rapidly. For instance, only about 0.1% of the data points exceed $\bar{p} + 3\sigma$, and this can be used to justify the following well-known criterion for detecting an outlier:

> If a data point is z (typically $z = 3$) or more standard deviations away from the arithmetic mean, then it is an outlier.

Such approaches, based on the normal distribution, have existed for many years. In 1863, Chauvenet [22] proposed to represent p as $p = \bar{p} + z\sigma$; then the normal distribution function is first used to determine the probability π that a data point $\in \mathscr{D}$ has a value $\geq \bar{p} + z\sigma$, and p is considered to be an outlier if $\pi |\mathscr{D}| < 0.5$ (the threshold of 0.5 is somewhat arbitrary). A more rigorous criterion, but essentially the

© Springer International Publishing AG 2017

K.G. Mehrotra et al., *Anomaly Detection Principles and Algorithms*, Terrorism, Security, and Computation, https://doi.org/10.1007/978-3-319-67526-8_6

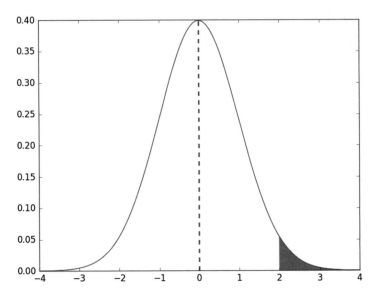

Fig. 6.1 Gaussian distribution with mean 0 and standard deviation = 1; probability in the right tail (*red area*) is 0.0227

same as described above, was earlier developed by Peirce in 1852 [48, 93]. Peirce's approach first computes a table estimating the maximum allowable deviation from the mean, beyond which a point would be considered an outlier. Values in the table depend on the size of the data set. If an outlier is found, subsequent iterations are initiated, looking for additional outliers.

Grubbs [49] argues that z should depend on the size $N = |\mathcal{D}|$ of the data set, especially when $|\mathcal{D}|$ is small, as follows. In his approach the goal is to determine if $p \in \mathcal{D}$ is an outlier with a desired degree of confidence $= 1 - \alpha$, provided:

$$\frac{|p - \bar{p}|}{\sigma} = z \geq \frac{N-1}{\sqrt{N}} \sqrt{\frac{t^2_{\frac{\alpha}{2N},N-1}}{N - 2 + t^2_{\frac{\alpha}{2N},N-1}}}.$$

Typically, $\alpha = 0.05$ or 0.01 (Fig. 6.1).

From the perspective of clustering, the assumption is that there is only one cluster in the dataset, and in it points are symmetrically distributed. Even in such data sets, measuring the distance from the mean is not always satisfactory because the mean of a set of observations is significantly skewed in the presence of an outlier. The notion of "distance from the rest of the data" may then be better captured in terms of the change in variance when a potential outlier is eliminated.[1] If an observation is an outlier, then its contribution to the sample variance will be large. Using this idea,

---

[1] The change in variance is also closely related to the change in the information content, e.g., the number of bits needed to describe $\mathcal{D} \setminus \{p\}$ vs. $\mathcal{D}$.

Grubbs proposed an alternative algorithm that computes the relative anomalousness of a data point as the ratio of sample variance *without* the largest observation with the sample variance *with* the largest observation to decide if the largest observation is anomalous. According to this criterion, p is an anomalous observation if

$$\sigma^2(\mathscr{D} \setminus \{p\})/\sigma^2(\mathscr{D})$$

is "small'. This approach can be extended to test if multiple data points are outliers, e.g., both p_1 and p_2 are considered anomalous if $\sigma^2(\mathscr{D} \setminus \{p_1, p_2\})/\sigma^2(\mathscr{D})$ is small.

Extensions of the above approaches for multi-dimensional datasets are straightforward:

- We may compute the anomalousness of a point separately along each dimension of the data, and aggregate the results. One possible rule is to use the maximum, that is, if $\alpha_i(p)$ represents the anomalousness of point p in the ith dimension, then $\alpha(p) = \max_i(\alpha_i(p))$, emphasizing points that are anomalous along any dimension.
- We may instead choose to consider anomalous points as those which are anomalous along many dimensions, captured if we define $\alpha(p) = \sum_i \alpha_i(p)$.
- Alternatively, we may reduce d-dimensional observations to scalars, e.g., using the Mahalanobis transformation $y = (p - \bar{p})^T S^{-1}(p - \bar{p})$ and then evaluate the (relative) anomalousness of y values. Note that for large datasets the values of y satisfy a χ^2 distribution if the original data (p) exhibit a multivariate Gaussian distribution.

Such simple tests are based on an implicit assumption that the data belongs to one cluster and satisfies the Gaussian distribution. If dataset $\mathscr{D}$ can be described as consisting of multiple well-separated clusters, then the distance-based criteria can be applied while restricting attention to the specific clusters to which different points belong. First, we must identify the nearest cluster $C_j \subset \mathscr{D}$ to which the point p belongs, e.g., based on minimal distance to the cluster centroid. We then consider the anomalousness of p with respect to C_j, using the approaches discussed earlier, and ignore other subsets of $\mathscr{D}$. For instance, if p belongs to cluster C_j and in ith dimension $\bar{p}_{j,i}$ denotes cluster centroid and σ_i the standard deviation, then the anomalousness of a point $p \subset \mathscr{D}$ is evaluated using the following expression:

$$\alpha(p) = \max_{i=1,\dots,d} \frac{|p_i - \bar{p}_{j,i}|}{k\sigma_i}.$$

If clusters are not clearly separated then the following approach is suitable. Suppose that $\mathscr{D} = C_1 \cup C_2 \cup \dots \cup C_k$, then $\alpha_{\mathscr{D}}(p) = \min_i \alpha_{C_i}(p)$.

The clustering or partitioning task may itself be described as an optimization task that minimizes the anomalousness of all data points, e.g., finding the choices of $C_1, \dots, C_k$ that will minimize

$$\sum_{p_j \in \mathscr{D}} \min_{i=1}^{k} \alpha_{C_i}(p_j).$$

6.1.1 Distance Based-Outlier Approach

Knorr and Ng [76] define the notion of distance based (DB) outliers "DB-outliers"
as follows: "an object p in a data set $\mathcal{D}$ is a DB(π, r)-outlier if at least a fraction
$\pi\mathcal{D}$ (where $0 \leq \pi \leq 1$) of the objects in $\mathcal{D}$ are at a distance greater than r from
p." where a distance r and a fraction π are two user selected parameters. In other
words, if we define the "r-neighborhood" of p as

$$\mathcal{N}_p(r) = \{q \; : q \in \mathcal{D} \text{ and } d(p, q) \leq r\},$$

then p is considered an outlier if

$$|\mathcal{N}_p(r)| \leq (1 - \pi)|\mathcal{D}|.$$

Knorr and Ng [75, 76] have shown that this notion of distance based outlier
detection generalizes the notion of outliers supported by statistical outlierness for
standard distributions. For example, DB$(0.9988, 0.13\sigma)$-outlierness corresponds to
the criterion $|p - \bar{p}| > 3\sigma$, for a normal distribution with mean $\bar{p}$ and standard
deviation σ. For a detailed description of this algorithm, see Algorithm "DB-
outlier".

Algorithm DB-outlier

Require: $\pi, r, \mathcal{D}$.
Ensure: List of outliers.

  1: $O = \emptyset$.
  2: **for** $p \in \mathcal{D}$ **do**
  3:     $N_p(r) = NULL$
  4:     **for** $q \in \mathcal{D}$ **do**
  5:         **if** $dist(p, q) <= r$ **then**
  6:             Insert q in $N_p(r)$
  7:         **end if**
  8:     **end for**
  9:     **if** $|N_p(r)| <= (1 - \pi)|\mathcal{D}|$ **then**
 10:         Insert p into O
 11:     **end if**
 12: **end for**

A straightforward implementation of the above criterion to determine all anomalous observations in $\mathcal{D}$ is "clearly" time consuming. To reduce the amount of
computation in the above algorithm, one approach is to identify non-anomalous
points as soon as possible. Most data points can be quickly identified as non-outliers
in case it can be determined that each of them is sufficiently near a non-trivial
fraction of the data set.

One approach to improving efficiency is called the Index-Based Algorithm, and
proceeds by counting the number of points within a distance of r from the point
p. As soon as this number exceeds $(1 - \pi) \times N + 1$, the point is declared to be a

non-outlier. If a standard multidimensional indexing structure (e.g., as described in Safar and Shahabi [102]) is used, then the complexity of this algorithm is quadratic in the size of the data set.

Another efficient approach is based on constructing a d-dimensional hyper-grid, as described below, where d is the dimensionality of the data points.

Consider a hyper-grid in d dimensions, where each side of a cell is of length $= \frac{r}{2\sqrt{d}}$, resulting in diagonal length of the cell equal to $r/2$. The following observations help in quickly determining non-anomalous observations:

- If two points lie within the same cell, the distance between them must be no larger than $r/2$.
- If two points lie in adjacent cells, referred to as "level-1 neighbors," then they may be arbitrarily close to each other, but the distance between them is guaranteed to be $< r$, twice the diagonal-length.
- "Level-2 neighbors" are in cells separated by exactly one cell; the minimum distance between level-2 neighbors is $\geq \frac{r}{2\sqrt{d}}$, the side-length, and the maximum distance between level-2 neighbors is $< 3r/2$, thrice the diagonal-length, hence $> r$.

Now consider the following setting:

- Let p be a point in cell C.
- Let cell C contain a points.
- Let b_j be the number of points in level j-neighboring cells, for $j = 1, 2$.

Then:

1. $a + b_1$ is a lower bound on $|\mathcal{N}_p(r)|$, hence p is not a DB-outlier if $a + b_1 > \lceil \pi |\mathcal{D}| \rceil$.
2. On the other hand, $a + b_1 + b_2$ is an upper bound on $|\mathcal{N}_p(r)|$, hence p is guaranteed to be a DB-outlier if $a + b_1 + b_2 < \lceil \pi |\mathcal{D}| \rceil + 1$.
3. If neither of the above conditions is satisfied, the point may or may not be an outlier, and additional analysis is required; for large data sets, the relative number of such points is expected to be small, so that the above tests are sufficient for most points, enabling faster computation of DB-outlierness.

Knorr and Ng have used the number of points within a fixed distance, r, from $p \in \mathcal{D}$ to evaluate its outlierness; the algorithms in later sections are motivated by the following criticisms of this approach:

- Several researchers have argued that this measure (number of points within r-distance) is inefficient in detecting anomalies in more complicated settings, particularly when $\mathcal{D}$ consists of clusters with varying densities. They propose that such counts should be measured at multiple distances and levels (within a small neighborhood).
- Another concern that has been raised about Knorr and Ng's approach is that a user may not know what to expect in a specific data set with reference to anomalies. In other words, the best choices of values for parameters (such as π and r) are hard to determine a priori.

- The size of the data set influences the results, whereas it may be argued that adding a collection of points at a great distance from p should not affect the outlierness of p.

The algorithms discussed in subsequent sections address these shortcomings.

6.2   Local Correlation Integral (LOCI) Algorithm

Papadimitriou et al. [92] propose a multi-point evaluation approach, called *Local Correlation Integral*, described below.

Let $p \in \mathcal{D}$ be an observation. As before, let $\mathcal{N}_p(r)$ be the set of all points in $\mathcal{D}$ within a distance of r from p; this set is also referred to as the *sampling neighborhood* of p. Given a predetermined parameter $0 < \alpha < 1$, define the *counting neighborhood* of p as $\mathcal{N}_p(\alpha r)$, the set of points within αr distance from p. Denote the number of points in this set as $n(p, \alpha r)$. Finally, denote by $\hat{n}(p, r, \alpha)$, the average, over all $\{q : q \in \mathcal{N}_p(r)\}$, of $n(q, \alpha r)$, i.e.,

$$\hat{n}(p, r, \alpha) = \frac{1}{|\mathcal{N}_p(r)|} \sum_{q \in \mathcal{N}_p(r)} n(q, \alpha r).$$

Given r and α, the *Multi-granularity Deviation Factor (MDEF)* at observation $p \in \mathcal{D}$ is defined as:

$$\text{MDEF}(p, r, \alpha) = 1 - \frac{n(p.\alpha r)}{\hat{n}(p, r, \alpha)}.$$

MDEF can be negative as well as positive. A negative value suggests that p is not anomalous, whereas a high value of MDEF suggests that p has relatively few near-neighbors, when compared to other points in the same region, hence p is more likely to be an anomaly.

These concepts are illustrated for a simple example in Fig. 6.2. Here, the r-neighborhood of p_0 contains 3 other observations, p_1, p_2, and p_3 (excluding itself). But none of these points are contained in the smaller αr-neighborhood. By contrast, the αr-neighborhoods of p_1, p_2, and p_3 contain 7, 3, and 6 observations respectively. Consequently, $\hat{n}(p, r, \alpha) = \frac{1+7+3+6}{4} = 4.25$, and $\text{MDEF}(p_0, r, \alpha) = 1 - 1/4.25 = 0.765$.

In order to determine the outlierness of the observation p, the LOCI algorithm proceeds as follows:

- The range of r values of interest is determined, with $r_{\max}$ chosen to be $\approx \alpha^{-1} \max_{p, q \in \mathcal{D}} \delta(p, q)$, and $r_{\min}$ is chosen so that the relevant neighborhoods contain approximately 20 observations.
- $\text{MDEF}(p, r, \alpha)$ values are computed for all $r \in [r_{\min}, r_{\max}]$.

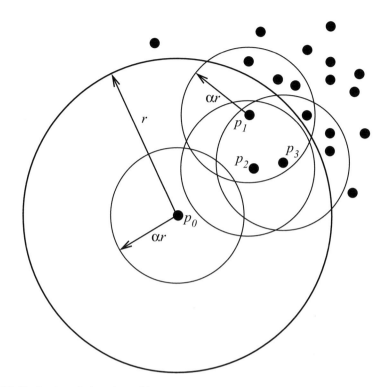

Fig. 6.2 Evaluation of $n(p, \alpha r)$ and $\hat{n}(p, r, \alpha)$ for a small dataset

- Let $\sigma_{\hat{n}}(p, r, \alpha)$ be the standard deviation of $n(q, \alpha r)$ values for $q \in \mathcal{N}_p(r)$, and

$$\sigma_{\text{MDEF}}(p, r, \alpha) = \frac{\sigma_{\hat{n}}(p, r, \alpha)}{\hat{n}(p, r, \alpha)}.$$

- An observation p is flagged as an outlier, if for any $r \in [r_{\min}, r_{\max}]$ its MDEF is sufficiently large. More precisely, p is considered to be an outlier, if

$$\text{MDEF}(p, r, \alpha) > k_\sigma \times \sigma_{\text{MDEF}}(p, r, \alpha).$$

Papadimitriou et al. suggest values of $\alpha = 1/2$ and $k_\sigma = 3$, although α can take any value between 0 and 1, and k_σ can also take any reasonable value.

At first glance, considering all values of $r \in [r_{\min}, r_{\max}]$ appears to be a daunting task but this is not a real problem because $n(p, r)$ and $n(p, r\alpha)$ (and therefore $\hat{n}(p, r, \alpha)$) change their values only a finite number of times, e.g., only when r increases enough to include at least one more observation in $\mathcal{N}_p(r)$ or $\mathcal{N}_p(\alpha r)$. Consequently, when $n(p, r)$ changes its value, values of r can be determined for all $p \in \mathcal{D}$.

Algorithm Exact LOCI algorithm

1: **for** each $p \in \mathscr{D}$ **do**
2:     Find $\mathcal{N}_p(r_{\max})$;
3:     Compute $\delta(p, p_{m\text{-}NN})$ and $\delta(p, \alpha p_{m\text{-}NN})$ for $1 \leq m \leq N$, where $p_{m\text{-}NN}$ denotes the mth nearest
        neighbor of p;
4:     Sort the list of these δs in ascending order of magnitude;
5:     For each r, in the sorted list, calculate $n(p, \alpha r)$ and $\hat{n}(p, \alpha r)$;
6:     Compute $\text{MDEF}(p, \alpha, r)$ and $\sigma_{\text{MDEF}}(p, \alpha, r)$;
7:     If $\text{MDEF}(p, \alpha, r) > 3 \sigma_{\text{MDEF}}(p, \alpha, r)$, then flag p as a potential outlier.
8: **end for**

The "exact LOCI" algorithm is presented in Algorithm "Exact LOCI algorithm". The time complexity of this exact algorithm can be reduced by resorting to some approximations [92].

6.2.1 Resolution-Based Outlier Detection

An alternative approach addresses the problem of parameter value determination by measuring the 'outlierness' of an observation $p \in \mathscr{D}$ at different resolutions, and aggregating the results [41].

In this algorithm, at the highest resolution, all observations are isolated points and thus considered to be outliers whereas at the lowest resolution all observations belong to one cluster and none is considered to be an outlier. As the resolution decreases from its highest value to the lowest value some observations in $\mathscr{D}$ begin to form clusters leaving other observations out of the clusters, and this phenomenon is captured in the resolution-based outlier detection approach.

The concept of resolution is related to the inter-observation distances. Naturally, if we consider $r_1 < \min_{i \neq j, p_i, p_j \in \mathscr{D}} d(p_i, p_j)$, then $\mathcal{N}_p(r_1)$ will contain only one point, p, for all $p \in \mathscr{D}$ and this represents the maximum resolution. On the other hand, if r^* (e.g., $r^* = \max_{p,q \in \mathscr{D}} d(p, q)$) is such that all observations in $\mathscr{D}$ belong to one cluster, then r^* corresponds to the smallest resolution.[2]

In the resolution-based approach, each cluster is formed using the transitive closure of the *close neighbor* relation. For $p, q \in \mathscr{D}$ and $r > 0$, the observation q is a close neighbor of p (with respect to r) if $d(p, q) \leq r$. The iterative accumulation of close neighbors results in a cluster, i.e., cluster growth continues until all close neighbors of points in a cluster are contained in the cluster.

Given r_1 and r^*, intermediate resolutions can be described by choosing $r_2 < r_3 < \ldots < r_R = r^*$. An example of this is equal spacing by $\Delta_R = (r^* - r_1)/R$, so that $r_i = r_{i-1} + \Delta_R$. Then the resolution-based outlier factor (ROF) is defined as:

---

[2]Fan et al. [41] consider the closeness of a point based on each dimension separately. Thus, in their implementation, an observation p is close to another observation q if the difference between p and q is less than r in any dimension. The algorithm is easier to implement using this definition of closeness since it avoids inter-observation distance computations.

Table 6.1 Illustration of ROF evaluation for observations in Fig. 6.3

Obs.	r_1 Cluster size	ROF	r_2 Cluster size	ROF	r_3 Cluster size	ROF	r_4 Cluster size	ROF
1	1	0	4	0	5	3/5	6	3/5+4/6
2	1	0	4	0	5	3/5	6	3/5+4/6
3	1	0	4	0	5	3/5	6	3/5+4/6
4	1	0	4	0	5	3/5	6	3/5+4/6
5	1	0	1	0	5	0	6	0+4/6
6	1	0	1	0	1	0	6	0+0

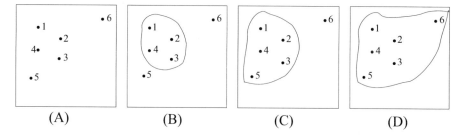

Fig. 6.3 Clustering of six observations based on successive decrease in resolution values

$$\mathrm{ROF}(p) = \sum_{i=1}^{R} \frac{\text{cluster-size}(p, r_{i-1}) - 1}{\text{cluster-size}(p, r_i)},$$

where cluster-size(p, r_i) represents the number of observations in the cluster that contains the observation $p \in \mathscr{D}$.

ROF evaluation is illustrated in Table 6.1 for a small dataset in Fig. 6.3. In this illustration, as expected, the smallest ROF corresponds to observation 6, which is the most significant outlier in this dataset.

In general, a small value of ROF(p) occurs when p belongs to fewer clusters, i.e., observations that join a cluster in a later iteration have smaller ROF values. The smallest ROF values correspond to the most anomalous observations.

Although ROF does not explicitly compute densities, the densest clusters will be formed first, followed by less dense clusters. Thus ROF incorporates some of the features of density based approaches, described in the following section.

6.3 Nearest Neighbor Approach

In the preceding sections, the numbers of observations within a fixed radius were used to determine a measure of outlierness. Alternatively, it can be argued that an observation is an outlier if its neighbors are far away. That is, a measure of outlierness can be described based on the distance of its neighbors.

In Ramaswamy et al. [98], the k-NN distance of a point is used as its outlier score. Let us denote the kth nearest neighbor of an observation p as $NN(p, k)$ and calculate $d(p, NN(p, k)) = d_k(p)$. A large value of $d_k(p)$ indicates that p is anomalous. The $d_k(p)$ values are ranked in decreasing order of magnitude and the points corresponding to the highest values are declared to be outliers. A detailed description of this algorithm is presented in Algorithm "k-NN outlier".

Algorithm k-NN outlier

Require: $k, n, \mathcal{D}$.
Ensure: O has n of outliers.

1: $O = \emptyset$.
2: **for** $p \in \mathcal{D}$ **do**
3:    $N_p = NULL$
4:    **for** $q \in \mathcal{D}$ **do**
5:       **if** $|N_p| < k$ **then**
6:          Add q in N_p
7:       **else**
8:          **if** $max\{dist(p, s)|s \in N_p\} > dist(p, q)$ **then**
9:             Add q in N_p and remove the first s in N_p such that $dist(p, s) > dist(p, q)$
10:          **end if**
11:       **end if**
12:    **end for**
13:    $NN(p, k) = max\{dist(p, s)|s \in N_p\}$
14: **end for**
15: **for** $p \in \mathcal{D}$ **do**
16:    **if** $|O| < k$ **then**
17:       Add p in O
18:    **else**
19:       **if** $min\{NN(s, k)|s \in O\} < NN(p, k)$ **then**
20:          Add p in O and remove the first s in O if $NN(s, k) < NN(p, k)$
21:       **end if**
22:    **end if**
23: **end for**
24: **return** O

Angiulli and Pizzuti [7] argue that the outlier criterion of Ramaswamy et al., does not provide good results, especially when $\mathcal{D}$ contains multiple clusters of different densities. They propose that a proper measure of outlierness is the *weight of p* defined as

$$\sum_{i=1}^{k} d_i(p).$$

As before, these weight values are sorted in decreasing order of magnitude to find the most anomalous points.

6.4   Density Based Approaches

Knorr and Ng's [76] approach fails to capture some outliers as illustrated in Fig. 6.4; here, if the object 'a' is considered as an DB(π, r)-outlier, then according to this definition all objects in cluster C_2 are also considered as DB(π, r)-outliers, which is counter-intuitive; by visual inspection, object 'a' appears to be more of an outlier than all objects in cluster C_2.

   To overcome such a deficiency, Breunig et al. [16], Tang et al. [108], Jin et al. [65] and others have suggested density based algorithms. The underlying assumption of a distance based anomaly detection algorithm is that the relative distances between other points in a neighborhood are irrelevant to anomaly detection; hence such an approach is assured to work well only when the different neighborhoods (populated by points) are characterized by roughly equal densities. But this assumption is often violated, i.e., in many practical situations data points have different densities in different neighborhoods. A density based approach looks at the "local" density of a point and compares it with density associated with its neighbors.[3]

   In density based approaches the main idea is to consider the behaviors of a point with respect to its neighbors' density values. The neighborhood is conceptualized

Fig. 6.4   An example in which DB(*pct*, dmin)-outlier definition and distance-based method do not work well. There are two different clusters C1 and C2, and one isolated data object 'a' in this data set. The distance from object 'b' to its nearest neighbor, d_2, is larger than the distance from object 'a' to its nearest neighbor, d_1, preventing 'a' from being identified as an anomaly

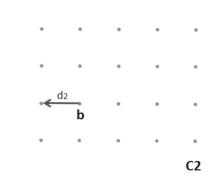

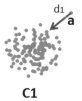

<hr />

[3]In the Syracuse region, a daily commuting distance characterized by a driving time of 25 min would be considered excessive; that same travel time would be considered to be low in the Los Angeles region. The definition of what constitutes "excessive driving time" must hence be a function of the distribution of driving times within the region of interest, rather than a constant over all regions.

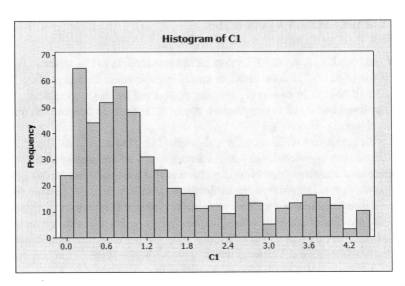

Fig. 6.5 Anomaly detection in one dimension using a histogram. Bins centered at 3.0 and 4.2 consist of very few observations and therefore the associated observations may be considered anomalous

by considering k nearest neighbors, where k is either iteratively estimated or is a preassigned integer. The underlying assumption is that if the density at a point p is 'smaller' than the densities of its neighbors, it must be an anomaly. The main difference between the approaches described below is in how they define the 'local' behavior and related density. Density may be computed in many ways, some of which are computationally expensive.

When the data is one-dimensional, we can use histograms to estimate the density function of the entire dataset, and hence detect anomalies. The data space is first subdivided into various disjoint bins of equal sizes. Some bins are heavily populated, and points in these bins are not considered to be outliers. Other bins may contain a relatively small number of points, compared to nearby bins, and can hence be argued to be anomalous, as illustrated in Fig. 6.5. A hierarchical binning approach can also be used; points in a smaller bin of low density may be considered to be anomalous if the larger bin (containing this smaller bin) is of higher density. The histogram method does not care where the bins of small densities are located; this allows for the possibility that the data distribution is not Gaussian, and that the dataset contains multiple clusters; it is also possible to discover anomalies that are not at the extremities of the data distribution, e.g., between two clusters.

For multi-dimensional data, most algorithms rely on estimating density from the statistics of the distances within a region of the data space, i.e., if average distances (between all points in a region) are small, then the density (in that region) is high, and conversely if the distances are large, the density is low. One concrete formulation of this approach is as follows:

- The local density of a point p is defined as the reciprocal of the average distance among the k points nearest to p.
- The relative anomalousness or outlier 'score' of p varies inversely with the local density at p.

Thus, one can calculate the outlier score of each point in the dataset, and sort the scores in decreasing order of magnitude; points at the top of the sorted list and with significantly 'large' scores are declared outliers.

6.4.1 Mixture Density Estimation

Often data is generated by more than one process, each characterized by a different distribution. In that case we may be able to model the data using a mixture of distributions and estimate the density at a point. The density estimation problem can be loosely defined as follows: given dataset $\mathcal{D}$ and a family of probability density functions, find the probability density that is most likely to have generated the given points. For concreteness, if the family of densities is multivariate Gaussian and we suspect that the data in $\mathcal{D}$ is generated by k different Gaussian distributions with density functions $g(p; \mu_j, \Sigma_j)$ for $j = 1, \ldots, k$, then we wish to find the 'best' set of parameters $\theta = (\pi_j, \mu_j, \Sigma_j : 1 \leq j \leq k)$ such that

$$f(p; \theta) = \sum_{1 \leq j \leq k} \pi_j g(p; \mu_j, \Sigma_j)$$

is most likely to have generated the given dataset. Of course, the parameters $(\pi_j, \mu_j, \Sigma_j : 1 \leq j \leq k)$ are unknown. One well-known approach to find the 'best' $f(p; \theta)$ is via Maximum Likelihood approach [31, 32]. After estimating the parameters of this mixture model, the anomalousness of an observation can be determined by the magnitude of the density at the point in comparison with the density at other points.

Alternatively, one can use a nonparametric density estimation as outlined below. Given $\mathcal{D}$, estimate the density at a point p as

$$f(p) = \frac{1}{Nh} \sum_{q \in \mathcal{D}} K(\frac{p - q}{h})$$

where h is the smoothing parameter. This density estimate can be applied to determine anomalous of p as described above—that is, if $f(p)$ is large, then p is considered to be a normal observation, and an anomaly otherwise. In this approach the kernel K is chosen to satisfy the following conditions.

1. K is a probability density function,
2. Typically K is a symmetric function, i.e., $K(u) = K(-u)$.

The following kernel functions are frequently used for the density estimation task:

1. $K(u) = \frac{3}{4}(1 - u^2)$, for $-1 \leq u \leq 1$, 0 otherwise
2. $K(u) = \frac{35}{32}(1 - u^2)^3$, for $-1 \leq u \leq 1$, 0 otherwise
3. $K(u) = \frac{1}{\sqrt{2\pi}} \exp(-\frac{u^2}{2})$, $-\infty < u < \infty$

This simple algorithm compares surprisingly well with many other algorithms, although it requires considerable computation for large and high-dimensional data spaces.

Three well-known density-based algorithms are described in the following subsections:

- Breunig et al. [16] suggest using the local outlier factor (LOF);
- Tang et al. [108] obtain the connectivity-based outlier factor (COF);
- Jin et al. [65] assign to each object the degree of being influenced outlierness (INFLO) and introduce a new idea called 'reverse neighbors' of a data point when estimating its density distribution.

The common theme among these algorithms is that they all assign outlierness to each object in the data set and an object will be considered as an outlier if its outlierness is greater than a pre-defined threshold (usually the threshold is determined by users or domain experts).

6.4.2 Local Outlier Factor (LOF) Algorithm

Breunig et al. [16] proposed the following approach to find anomalies in a given dataset. As the name of the algorithm suggests, the Local Outlier Factor (LOF) measures the local deviation of a data point $p \in \mathscr{D}$ with respect to its k nearest neighbors. A point p is declared anomalous if its LOF is 'large.' The LOF of a point is obtained as described in the following steps:

LOF Computation

1. Find the distance, $d_k(p)$, between p and its kth nearest neighbor. The distance can be any measure, but typically the Euclidean distance is used.
2. Let the set of k nearest neighbors of p be denoted by $\mathcal{N}_k(p) = \{q \in \mathscr{D} - \{p\} : d(p, q) \leq d_k(p)\}$.
3. Define the reachability distance of a point q from p, as $d_{reach}(p, q) = \max\{d_k(q), d(p, q)\}$. This is illustrated in Fig. 6.6.
4. The average reachability distance of p is

$$\overline{d_{reach}}(p) = \frac{\sum_{q \in \mathcal{N}(p)} d_{reach}(p, q)}{|\mathcal{N}_k(p)|}.$$

Fig. 6.6 Illustration of
reachability distance.
$d_{reach}(p1, o)$ and $d_{reach}(p2, o)$
for $k = 4$

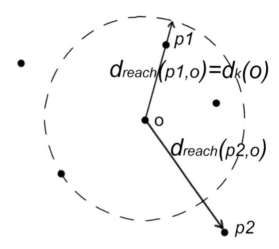

The local reachability density of a point is defined as the reciprocal of reachability distance

$$\ell_k(p) = [\overline{d_{reach}(p)}]^{-1}.$$

5. Finally, this local reachability density is compared with the local reachability densities of all points in $\mathcal{N}_k(p)$, and the ratio is defined as LOF (local outlier factor):

$$\mathcal{L}_k(p) = \left[\frac{\sum_{o \in \mathcal{N}_k(p)} \frac{\ell_k(o)}{\ell_k(p)}}{|\mathcal{N}_k(p)|} \right].$$

6. The LOF of each point is calculated, and points are sorted in decreasing order of $\mathcal{L}_k(p)$. If the LOF values are 'large', the corresponding points are declared as outliers.
7. To account for k, the final decision is taken as follows: $\mathcal{L}_k(p)$ is calculated for selected values of k in a pre-specified range, max $\mathcal{L}_k(p)$ is retained, and a p with large LOF is declared an outlier.

A detailed description of this algorithm is presented in on the next page (Algorithm "LOF Computation").

A variation of LOF is Cluster-based LOF (CBLOF), proposed by Gao [47]. In this case the first step is to find clusters of the dataset and divide all clusters into two broad groups—small clusters and large clusters. The CBLOF of each point in $\mathcal{D}$ is obtained depending on the data point belong to a small cluster or a large cluster. If the data point lies in a 'small' cluster, then its CBLOF score is obtained as the product of the size of the cluster and the point's distance to the nearest centroid of a large cluster. However, if the point belongs to a large cluster, then its score is the product of the size of the cluster and point's distance from the cluster's centroid. Other extensions, such as the local density based outlier factor have also been proposed.

Algorithm LOF (Local Outlier Factor) Computation

Require: $k, \mathcal{D}$.
Ensure: L_k - LOF score for each object in $\mathcal{D}$

1: $L_k = \emptyset$.
2: **for** $p \in \mathcal{D}$ **do**
3:    $N_k(p) = NULL$
4:    **for** $q \in \mathcal{D}$ **do**
5:      **if** $|N_k(p)| < k$ **then**
6:        Add q in $N_k(p)$
7:      **else**
8:        Let $s* \in N_k(p)$ be such that $dist(p, s*) \geq dist(p, s)$ for all $s \in N_k(p)$;
9:        **if** $dist(p, s*) > dist(p, q)$ **then**
10:          Replace $s* \in N_k(p)$ by q
11:        **end if**
12:      **end if**
13:    **end for**
14:    $d_k(p) = \max\{dist(p, s) | s \in N_k(p)\}$
15: **end for**
16: **for** $p \in \mathcal{D}$ **do**
17:    **for** $q \in \mathcal{D}$ **do**
18:      $d_{reach}(p, q) = \max\{d_k(p), d(p, q)\}$
19:    **end for**
20: **end for**
21: **for** $p \in \mathcal{D}$ **do**
22:    $l_k(p) = \frac{|N_k(p)|}{\sum_{q \in N_k(p)} d_{reach}(p, q)}$
23: **end for**
24: **for** $p \in \mathcal{D}$ **do**
25:    $L_k(p) = \left[\frac{\sum_{o \in N_k(p)} \frac{l_k(o)}{l_k(p)}}{|N_k(p)|}\right]$
26: **end for**
27: **return** L_k

6.4.3 Connectivity-Based Outlier Factor (COF) Approach

LOF performs well in many application domains, but its effectiveness will diminish if the density of an outlier is close to densities of its neighbors. For example, in Fig. 6.7, data object $o1$ is isolated but its density is very close to densities of objects in cluster 1, $C1$, hence LOF fails to detect this outlier.

To solve such a deficiency of LOF, Tang et al.[107] suggest a new method to calculate the density as described below. Define the distance between two non-empty sets P and Q as $\mathsf{d}(P, Q) = \min\{\mathsf{d}(p, q) : p \in P, q \in Q\}$. This can be used to define the minimum distance between a point and a set by treating the point as a singleton set.

1. As in the previous algorithm, let $\mathcal{N}_k(p)$ be the set of k nearest neighbors of p.
2. The "set-based path (SBN)" is an ordered list of all neighbors of p, arranged in increasing order of distance from p. Formally, the SBN of length k is a path $< p_1, p_2, \ldots, p_k >$ based on the set $\{p, \mathcal{N}_k(p)\}$ such that for all $1 \leq i \leq k - 1$, p_{i+1} is the nearest neighbor of the set $\{p_1, p_2, \ldots, p_i\}$ in $\{p_{i+1}, p_{i+2}, \ldots, p_k\}$.

Fig. 6.7 An example in which LOF fails for outlier detection

Fig. 6.8 Illustration of set-based path and trail definitions. The k nearest neighbors of p are $q1, q2$, and $q3$, for $k = 3$. Then SBN path from p is $\{p, q1, q3, q2\}$, and SBT is $< e1, e2, e3 >$ respectively

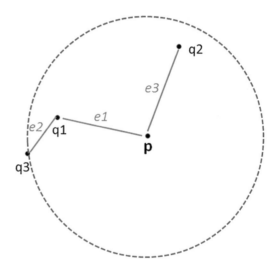

3. The Set-based trail (SBT) is an ordered collection of $k - 1$ edges associated with a given SBN path $< p_1, p_2, \ldots, p_k >$. The ith edge e_i connects a point $o \in \{p_1, \ldots, p_i\}$ to p_{i+1} and is of minimum distance; i.e., length of e_i is

$$|e_i| = d(o, p_{i+1}) = d(\{p_1, \ldots, p_i\}, \{p_{i+1}, \ldots, p_k\}).$$

   Figure 6.8 illustrates these concepts.

4. Given p, the associated SBN path $< p \equiv p_1, p_2, \ldots, p_k >$, and the SBT $< e_1, e_2, \ldots, e_{k-1} >$, the weight w_i for edge e_i is proportional to the order in which it is added to SBT set. Then the average-chaining distance ($\mathcal{A}$) of p is the weighted sum of the lengths of the edges. That is:

$$\mathcal{A}_{\mathcal{N}_k(p)}(p) = \sum_{i=1}^{k-1} w_i \times |e_i|.$$

where

$$w_i = \frac{2(k-i)}{k(k-1)}$$

5. Finally, the connectivity-based outlier factor (COF) of a point p is defined as the ratio of p's average-chaining distance with the average of average-chaining distances of its k nearest neighbors;

$$\text{COF}_k(p) = [A_{\mathcal{N}_k(p)}(p)] \left[\frac{\sum_{o \in \mathcal{N}_k(p)} A_{\mathcal{N}_k(p)}(o)}{|\mathcal{N}_k(p)|} \right]^{-1}.$$

6. Larger values of $\text{COF}_k(p)$ indicates that p is more anomalous.

COF works better than LOF in the data sets with sparse neighborhoods (such as a straight line), but its computation cost is higher than LOF. A detailed description of this algorithm is presented on the next page.

6.4.4 *INFLuential Measure of Outlierness by Symmetric Relationship (INFLO)*

This approach, proposed by Jin et al. [65], uses the notion of "reverse nearest neighbors" of an object p to obtain a measure of outlierness. The main idea is that an object is an outlier if it is not a nearest neighbor of its own nearest neighbors.

The "Reverse Nearest Neighborhood (RNN)" of an object p is defined as

$$\mathcal{RN}_k(p) = \{q : q \in \mathcal{D} \text{ and } p \in \mathcal{N}_k(q)\}.$$

Note that $\mathcal{N}_k(p)$ has k objects but $\mathcal{RN}_k(p)$ may not have k objects. In some instances, it may be empty, because p may not be in any of $\mathcal{N}_k(q)$ for any $q \in \mathcal{N}_k(p)$ For example, consider the following case:

- The two points nearest to p are q_1, and q_2.
- The two nearest neighbors of q_1 are q_2, and q_3,
- The two nearest neighbors of q_2 are q_1 and p.

Then, for $k = 2$, the set of reverse nearest neighbors of p is $\{q_2\}$ (Fig. 6.9).

Jin et al. [65] propose to replace $\mathcal{N}_k(p)$ by the k-influential space for p, denoted as $\text{IS}_k(p) = \mathcal{N}_k(p) \cup \mathcal{RN}_k(p)$. Associated with $\text{IS}_k(p)$, they define the influential outlierness of a point p as

$$\text{INFLO}_k(p) = \frac{1}{\text{den}(p)} \frac{\sum_{o \in \text{IS}_k(p)} \text{den}(o)}{|(\text{IS}_k(p))|}$$

Fig. 6.9 Reverse Nearest
Neighborhood and Influence
Space. For $k=3$, $\mathcal{N}_k(q_5)$ is
$\{q_1, q_2, q_3\}$.
$\mathcal{N}_k(q_1) = \{p, q_2, q_4\}$.
$\mathcal{N}_k(q_2) = \{p, q_1, q_3\}$.
$\mathcal{N}_k(q_3) = \{q_1, q_2, q_5\}$.
$\mathcal{R}\mathcal{N}_k(q_5) = \{q_3, q_4\}$

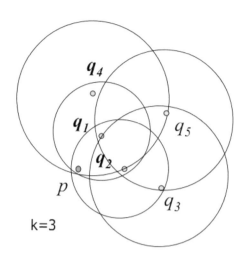

Algorithm COF

Require: $k, \mathcal{D}$.
Ensure: COF_k - COF score for each object in $\mathcal{D}$

1: $COF_k = \emptyset$.
2: **for** $p \in \mathcal{D}$ **do**
3:     $N_k(p) = NULL; SBN(P) = \{P\}; SBTDist(p) = \emptyset; A_{N_k}(p) = 0$
4:     **for** $q \in \mathcal{D}$ **do**
5:         **if** $|N_k(p)| < k$ **then**
6:             Add q in $N_k(p)$
7:         **else**
8:             Let $s* \in N_k(p)$ be such that $dist(p, s*) \geq dist(p, s)$ for all $s \in N_k(p)$;
9:             **if** $dist(p, s*) > dist(p, q)$ **then**
10:                 Replace $q* \in N_k(p)$ by q
11:             **end if**
12:         **end if**
13:     **end for**
14:     $i=1; NN(p) = N_k(p)$
15:     **while** $|NN(p)| > 0$ **do**
16:         $dist(e_i) = min\{dist(s, t)|s \in SBN(p), t \in NN(p)\}$
17:         Move corresponding t from $NN(p)$ to $SBN(p)$
18:         $i++$
19:         Add $dist(e_i)$ to $SBTDist(p)$
20:     **end while**
21:     **for** $i=1$ to k **do**
22:         $A_{N_k}(p) = A_{N_k}(p) + \frac{dist(e_i)*2(k+1-i)}{(k+1)*k}$
23:     **end for**
24: **end for**
25: **for** $p \in \mathcal{D}$ **do**
26:     $COF_k(p) = \frac{|N_k(p)|*A_{N_k}(p)}{\sum_{o \in N_k(p)} A_{N_k}(o)}$
27: **end for**
28: **return** COF_k

where $\text{den}(p) = \frac{1}{d_k(p)}$; $d_k(p)$ represents the distance between p and its kth nearest neighbor.

Thus for any p, INFLO expands $\mathcal{N}_k(p)$ to IS_k and compares p's density with average density of objects in IS_k. By using the reverse neighborhood, INFLO enhances its ability to identify the outliers in more complex situation, but its performance is poor if an object p's neighborhood includes data objects from groups of different densities, then outlierness of p cannot be correctly measured.

Algorithm INFLO

Require: $k, \mathcal{D}$.
Ensure: $INFLO_k$ - INFLO score for each object in $\mathcal{D}$

1: $INFLO_k = \emptyset$.
2: **for** $p \in \mathcal{D}$ **do**
3:     $N_k(p) = \emptyset; RN_k(p) = \emptyset$;
4:     **for** $q \in \mathcal{D}, q \neq p$ **do**
5:         **if** $|N_k(p)| < k$ **then**
6:             Add q in $N_k(p)$
7:         **else**
8:             Let $s* \in N_k(p)$ be such that $dist(p, s*) \geq dist(p, s)$ for all $s \in N_k(p)$;
9:             **if** $dist(p, s*) > dist(p, q)$ **then**
10:                Replace $q* \in N_k(p)$ by q
11:            **end if**
12:        **end if**
13:    **end for**
14:    $d_k(p) = \max\{dist(p, s) | s \in N_k(p)\}$
15:    **for** $q \in N_k(p)$ **do**
16:        Add p in $RN_k(q)$
17:    **end for**
18: **end for**
19: **for** $p \in \mathcal{D}$ **do**
20:    $IS_k(p) = N_k(p) \cup RN_k(p)$
21:    $INFLO_k(p) = \dfrac{d_k(p) \cdot \sum_{o \in IS_k(p)} \frac{1}{d_k(o)}}{|IS_k(p)|}$
22: **end for**
23: **return** $INFLO_k$

6.5 Performance Comparisons

Algorithms discussed in this chapter were compared for multiple datasets, described in the Appendix, to evaluate their performance.

- LOF, COF, and INFLO were compared using the Synthetic dataset1 for $k = 4, 5, 6$, and 7, and $m_t = 5, 6$. It was observed that all three algorithms have identical performance in all cases except when $k = 4$ and $m_t = 6$. In this case, performance of COF was slightly better, as measured using RankPower which was 1 for COF and 0.955 for LOF and INFLO.

- LOCI, ALOCI, LOOP, INFLO, KNN, LOF, and COF were compared using Iris, KDD-1000, and PEC-1000 datasets and were compared in terms of their performance as measured in terms of RankPower.
  - For the Iris data, each of the algorithms found exactly five out of the ten anomalies.
  - For the KDD-1000 dataset, ALOCI and LOF show much better performance than the other algorithms. For example, these two algorithms find 9 and 8 anomalies out of 10 compared to other algorithms which identified only 1 or 2 anomalies correctly. In another experiment with the same datasets when 20 anomalies were inserted, LOCI, ALOCI, and LOF found 9 anomalies whereas the other algorithms found only 1 or 2 anomalies.
  - For the PEC-1000 the performance of LOCI, ALOCI, and KNN was only relatively better than the rest of the algorithms. For example, out of 10 inserted anomalies, these three algorithms identified 4, 5, and 6 anomalies correctly whereas the other algorithms found only 3 or less anomalies.

Based on these experiments, we conclude that ALOCI consistently performs better than other algorithms. Other algorithms, such as LOCI, LOF, and KNN also provide good performance in some datasets.

6.6   Conclusions

This chapter has presented several distance and density-based anomaly detection approaches and algorithms. These include algorithms based on direct distance computations, which do not perform well if the data set is characterized by variations in density. Other variations of distance-based approaches discussed in the literature include [38, 39, 121]. Algorithms based on density estimation include classical kernel-based methods as well as algorithms that use nearest neighbor distance computations to estimate density, sometimes indirectly.

Chapter 7
Rank Based Approaches

The density-based methodology discussed in the preceding chapter, which examines the k-neighborhood of a data point, has many good features. For instance, it is independent of the distribution of the data and is capable of detecting isolated objects. However it has the following shortcomings:

- If some neighbors of the point are located in one cluster, and the other neighbors are located in another cluster, and the two clusters have different densities, then comparing the density of the data point with all of its neighbors may lead to a wrong conclusion and the recognition of real outliers may fail, an example is illustrated in Sect. 7.1.
- The notion of density does not work well for sparse data sets such as a cluster of points on a single straight line. Even if each point in the set has equal distances to its closest neighbors, its density may vary depending on its position in the dataset.

Clusters with different densities and sizes arise in many real-world context. An example is the case where a financial institution wishes to find anomalous behavior of its customers. It is obvious that all normal customers are not alike—depending upon their needs and nature all have different 'normal' behavior. Individuals with multiple accounts and large investments tend to have deposit and withdrawal patterns that differ substantially from those of individuals with small amounts. Consequently, the collection of data associated with *all* users is likely to form multiple clusters with variable number of data points in each cluster, and with variable densities, etc. In such a situation, density based approaches will perform poorly.

In such situations, to find anomalous observations, the ideal solution is to transform the data so that all regions in the transformed space have similar local distributions. The rank-based approach attempts to achieve this goal. In using the rank based approach the effect of inter-object distances is diminished; and in using 'modified-ranks', described in Sect. 7.3, the effect of the size of the local cluster(s) is also accommodated.

© Springer International Publishing AG 2017
K.G. Mehrotra et al., *Anomaly Detection Principles and Algorithms*, Terrorism, Security, and Computation, https://doi.org/10.1007/978-3-319-67526-8_7

Fig. 7.1 Example of an outlier (Bob) identified by the fact that he is not very close to his best friend (Eric), unlike an inlier (Jack)

The idea of rank is borrowed from the research literature on social networks. Detecting outliers in a given data set resembles finding the most unpopular person in a given social network. Consider the social network in Fig. 7.1: to discover the relative popularity of Bob, we may ask Bob:"who are your k best friends?". Then we may ask all "friends" of Bob the same question. If Bob is not listed among the close friends of his friends, clearly he is not a popular person. A summary of answers from all k persons whom Bob identifies as his friends allows us to draw a clear conclusion about his popularity. Just as we can use the concept of relative popularity in friend networks, we can use a similar notion of rank to capture the outlierness of a point.

In this chapter new approaches based on a rank measure and clustering are presented for outlier detection.

7.1 Rank-Based Detection Algorithm (RBDA)

In this section, we consider a new approach to identify outliers based on mutual proximity of a data point and its neighbors. The key idea of the algorithms is to use rank instead of distance. These ranks are relative to a point $p \in \mathcal{D}$; that is, given a point $p \in \mathcal{D}$, the largest rank is assigned to the point $q \in \mathcal{D}$ that is farthest away from p and smallest rank, 1, to the point closest to it.

To understand mutual proximity consider $p \in \mathcal{D}$ and $q \in \mathcal{N}_p(k)$, where $\mathcal{N}_p(k)$ is the k-neighborhood of p. Here q is "close" to p because it belongs to $\mathcal{N}_p(k)$. In return, we ask "how close is p to q?". If p and q are 'close' to each other, then we argue that (with respect to each other) p and q are not anomalous data points. However, if q is a neighbor of p but p is not a neighbor of q, then with respect to q, p is an anomalous point. If p is an anomalous point with respect to most of its neighbors, then p should be declared to be an anomaly. When measuring the outlierness of p, instead of distance, we use the ranks calculated based on neighborhood relationships between p and $\mathcal{N}_k(p)$, see Fig. 7.2. Low cumulative rank assigned to p by its neighbors implies that p is a central point because it is among the nearest neighbors of its own closest neighbors. However a point at the periphery of a cluster has a high cumulative sum of ranks because its nearest neighbors are closer to each other than the point. This forms the basis of RBDA [59].

Description of Rank-Based Detection Algorithm (RBDA) Algorithm

1. Let $p \in \mathcal{D}$ and $\mathcal{N}_k(p)$ denotes the set of its k-neighbors. Let $r_q(p)$ denote the rank of p with respect to q, that is:

$$r_q(p) = \text{number of points } o \in \mathcal{D} - \{p\} \text{ such that } d(q,o) < d(p,q).$$

For all $q \in \mathcal{N}_k(p)$, calculate $r_q(p)$.

Fig. 7.2 Illustration of ranks: *Red dash* shows k-NN of p when k is 3 and *long blue dash* shows a circle with radius of $d(p, q3)$ and center of $q3$. The k-NN of p is $\{q1, q2, q3\}$. Then $r_{q3}(p)=4$, because $d(q3, p)$ is greater than any of $d(q3, q3), d(q3, q4), d(q3, q1)$ and $d(q3, q5)$

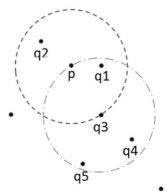

2. 'Outlierness' of p, denoted by $\mathcal{O}_k(p)$, is defined as:

$$\mathcal{O}_k(p) = \frac{\sum_{q \in \mathcal{N}_k(p)} r_q(p)}{|\mathcal{N}_k(p)|}. \qquad (7.1)$$

If $\mathcal{O}_k(p)$ is 'large' then p is considered an outlier.

3. One criterion to determine 'largeness' is described below. Let $D_o = \{p \in \mathcal{D} \mid \mathcal{O}_k(p) \leq \mathcal{O}^*\}$ where $\mathcal{O}^*$ is chosen such that the size of D_o is a large fraction (e.g. 75%) of the size of $\mathcal{D}$. We normalize $\mathcal{O}_k(p)$ as follows:

$$\mathcal{L}_k(p) = ln(\mathcal{O}_k(p)) \qquad (7.2)$$

$$Z_k(p) = \frac{1}{S_k}(\mathcal{L}_k(p) - \bar{\mathcal{L}}_k) \qquad (7.3)$$

where

$$\bar{\mathcal{L}}_k = \frac{1}{|D_o|} \sum_{p \in D} \mathcal{L}_k(p) \text{ and } S_k^2 = \frac{1}{|\mathcal{D}_o| - 1} \sum_{p \in \mathcal{D}} (\mathcal{L}_k(p) - \bar{\mathcal{L}}_k)^2$$

and if the normalized value $Z_k(p)$ is ≥ 2.5, then we declare that p is an outlier. In this criterion we have assumed that the distribution of $Z_k(p) = \frac{1}{S_k}(\mathcal{L}_k(p) - \bar{\mathcal{L}}_k)$ (normalized for mean and standard deviation) will be approximated by the standard normal random variable and $P(Z_k(p) = \frac{1}{S_k}(\mathcal{L}_k(p) - \bar{\mathcal{L}}_k) > 2.5) \approx 0.006$. Hence, value of $Z_k(p) = \frac{1}{S_k}(\mathcal{L}_k(p) - \bar{\mathcal{L}}_k) > 2.5$ will be an outlier.

4. Alternatively, one can sort observations in decreasing order of $\mathcal{O}_k(p)$ and declare the top few observations as anomalous. A histogram of $\mathcal{O}_k(p)$ values can be used for easy visualization.

A detailed description of the RBDA algorithm is presented in Algorithm "Rank-Based Detection Algorithm".

7.1.1   Why Does RBDA Work?

Before explaining why RBDA works, first we examine a scenario in which the density-based algorithm would fail. Consider the data set in Fig. 7.3. There are three groups of data objects and one isolated data object—'A'. Data object 'B' is from group3. When k is 7, the k nearest neighbors of both 'A' and 'B' contain the data objects from different density groups. In this case, the density-based outlier detection algorithm, LOF, assigns a higher outlierness value 1.5122 to 'B' and lower outlierness value 1.1477 to 'A' which is counter-intuitive. Density-based algorithms

---

Algorithm Rank-Based Detection Algorithm

Require: $k, \mathcal{D}$.
Ensure: $RBDA_k$ - RBDA score for each object in $\mathcal{D}$

1: $RBDA_k = \emptyset$.
2: **for** $p \in \mathcal{D}$ **do**
3:     $N(p) = \emptyset; N_k(p) = \emptyset;$
4:     **for** $q \in \mathcal{D}$ **do**
5:         **if** $q \neq p$ **then**
6:             Add q to $N(p)$
7:         **end if**
8:     **end for**
9:     Sort $N(p)$ by $dist(p, q)$ in ascending order
10:     $d_k(p)$=kth of $N(p)$
11:     $tmp = 0; index = 0; rank = 0;$
12:     **for** $q \in N(p)$ **do**
13:         **if** $dist(p, q) \leq d_k(p)$ **then**
14:             Add q in $N_k(p)$
15:         **end if**
16:         index++
17:         **if** $dist(p, q) \leq tmp$ **then**
18:             rank = index;
19:         **end if**
20:         $r_p(q) = rank; tmp = dist(p, q);$
21:     **end for**
22: **end for**
23: **for** $p \in \mathcal{D}$ **do**
24:     sumrank=0;
25:     **for** $q \in N_k(p)$ **do**
26:         sumranks += $r_q(p)$
27:     **end for**
28:     $RBDA_k(p) = \frac{sumranks}{|N_k(p)|}$
29: **end for**
30: **return** $RBDA_k$

---

assume that all neighbors of data object of interest are from the same density groups, but such is not the case in our example.[1]

To overcome this issue, instead of focusing on the calculation of density, RBDA chooses ranks instead of distance. Thus in Fig. 7.3, RBDA outlierness (average rank) of 'A' is 10 and that of 'B' is less (6.5714) as desired.

Use of ranks eliminates the problem of density calculation in the neighborhood of the point, and this improves performance. Although based on distance, ranks work better than distance and captures the anomalousness of an object more precisely in most cases. Consequently, rank-based methods tend to perform better than several density-based methods, on some synthetic data set as well as on some real datasets.

---

[1] In this illustration, LOF suffers the same deficiency when k is 6, 7, or 8. For example, LOF assigns outlierness value 1.5122 to 'B' and 1.1477 to 'A', and it fails to identify 'A' as the most significant outlier.

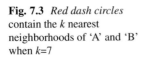

Fig. 7.3 *Red dash circles* contain the *k* nearest neighborhoods of 'A' and 'B' when *k*=7

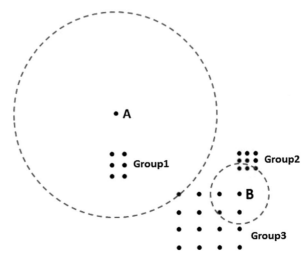

Rank based clustering can be used for detecting tight and sparse clusters easily using only one parameter. Based on ranks, it forms clusters with higher accuracy. By combining it with ranks, it achieves better performance than algorithms discussed in previous chapters and thus it is applicable in broad practical domains.

7.2 Anomaly Detection Algorithms Based on Clustering and Weighted Ranks

In Chap. 4 we argued that clustering can be used to find anomalous objects in a given data set. This approach has a significant advantage; it reduces the time complexity considerably provided the clustering algorithm is fast.

It can be seen that an object that is near a very large cluster is more likely to be declared anomalous even if it is only slightly far away from its neighbors, compared to the case if it had been close to a small cluster. To account for this imbalance, the concept of modified rank has been proposed.

In this section, we first describe a new clustering approach, a modification of DBSCAN, called Neighborhood Clustering (NC-clustering). Next we consider several algorithms to find anomalous observations in a dataset, where clustering is a preprocessing step. In a clustering based approach, observations that do not belong to a cluster are declared anomalous. In the following, this condition is relaxed; instead observations that do not belong to a cluster are *potential* anomaly candidates.

7.2.1 NC-Clustering

As in DBSCAN, in NC-clustering a cluster is required to have minimum number of objects, denoted as ℓ (in DBSCAN it is denoted as MinPts). The main difference between this approach and DBSCAN is in interpreting 'reachability'. In NC-clustering reachability is defined using k-neighborhood, i.e., the main idea is to declare data objects p and q in $\mathcal{D}$ to be close if q is one of the k-nearest neighbors of p and vice-versa. One distinct advantage of this approach is that k is easier to guess than the parameter 'eps' needed by DBSCAN to define the neighborhood size. Another difference is that, in DBSCAN, a border point is considered to be in the cluster; this is not the case in NC-clustering. Formally, the following definitions are used in the proposed clustering algorithm; all definitions are parameterized with a positive integer parameter ℓ intended to capture the notion of cluster tightness.

- *D-reachability* (given ℓ): An object p is directly reachable (D-reachable) from q, if $p \in \mathcal{N}_\ell(q)$.
- *Reachability*: An object p is reachable from q, if there is a chain of objects $p \equiv p_1, \ldots, p_n \equiv q$, such that p_i is D-reachable from p_{i+1} for all values of i.
- *Connectedness*: If p is reachable from q, and q is reachable from p, then p and q are connected.

Figure 7.4 illustrates the concepts of $d_k(p)$, $\mathcal{N}_k(p)$ and $\mathcal{RN}_k(p)$. In Fig. 7.4, for $k = \ell = 3$, y is D-reachable from p since y is in $\mathcal{N}_3(p)$, but p and y are not connected

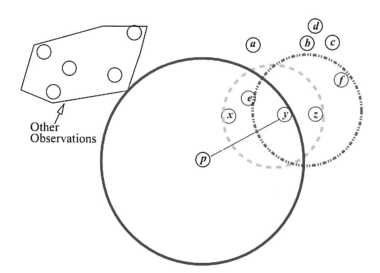

Fig. 7.4 Illustration of $d_k(p)$, $\mathcal{N}_k(p)$ and $\mathcal{RN}_k(p)$ for $k = 3$. The *large blue circle* contains elements of $\mathcal{N}_3(p) = \{x, e, y\}$. Because y is farthest third nearest neighbor of p, $d_3(p) = d(p, y)$, the distance between p and y. The *smallest green circle* contains elements of $\mathcal{N}_3(y)$ and the *red circle* contains elements of $\mathcal{N}_3(z)$. Note that $r_y(p) = $ rank of p among neighbors of $y = 9$. Finally, $\mathcal{RN}_k(y) = \emptyset$ since no other object considers p in their neighborhood

since p is not in $\mathcal{N}_3(y)$. However, y and z are connected since they are in each other's 3-neighborhoods. The clustering method is represented as $\mathcal{NC}(\ell, m^*)$.

Breadth-first search can be performed on a graph whose node-set is $\mathcal{D}$ and in which an edge exists between $p, q \in \mathcal{D}$ if $p \in \mathcal{N}_\ell(q)$ and $q \in \mathcal{N}_\ell(p)$. A connected component C of the graph is a cluster if the following conditions are satisfied:

1. Every object in C is D-reachable from at least two others in C.
2. The number of objects in C is no smaller than a pre-specified minimum, m^*.

For example, $\mathcal{NC}(6, 5)$ denotes that a cluster contains connected objects for $\ell = 6$, and every cluster must contain at least 5 objects. If any connected component C does not satisfy these conditions, it is broken up into isolated objects, which are declared to be potential outliers.

The appropriate values of ℓ and m^* are problem-specific and depend on domain knowledge. If ℓ is small, the NC-clustering method finds small and tightly connected clusters and large values of ℓ result in large and loose clusters. If the clusters are small and tight, more objects that do not belong to any cluster are detected and in the latter case only a few objects are declared as outliers. In real world applications (such as credit card fraud detection) most of the transactions are normal and only 0.01% or less of the transactions are fraudulent. In such cases, a small value of ℓ is more suitable than a large ℓ. The value of m^* has a similar effect: if m^* is too small, then the cluster size may also be too small, and a small collection of outliers may be considered as a cluster, which is not what we want.

Advantages of $\mathcal{NC}(\ell, m^*)$ clustering algorithms are:

- It only requires one scan to find all the clusters.
- It controls the tightness and sparsity of clusters using a single parameter—ℓ.
- It can find the central objects of clusters easily by analyzing each $\mathcal{O}_k(p)$. The data objects in the center must have lower $\mathcal{O}_k(p)$ values.

7.2.2 Density and Rank Based Detection Algorithms

In this section we consider algorithms that first use a clustering approach, such as the one described above, followed by methods that attempt to account for cluster sizes, and other relevant parameters.

Density-Based Clustering and Outlier Detection Algorithm (DBCOD)

For $p \in \mathcal{D}$, Tao and Pi [109] define the local density, the neighborhood-based density factor, and neighborhood-based local density factor of p, respectively, as:

$$\text{LD}_k(p) = \frac{\sum_{q \in \mathcal{N}_k(p)} \frac{1}{d(p,q)}}{|\mathcal{N}_k(p)|}, \quad \text{NDF}_k(p) = \frac{|\mathcal{RN}_k(p)|}{|\mathcal{N}_k(p)|}, \quad \text{and } \text{NLDF}_k(p) = \text{LD}_k(p) \times \text{NDF}_k(p).$$

The threshold of NLDF, denoted as τ_{NLDF}, is defined as:

$$\tau_{\text{NLDF}} = \begin{cases} \min_k(\text{NLDF}_k(p)) & \text{if for all objects } p \in D, \text{NDF}_k(p) = 1 \\ \max_k(\text{NLDF}_k(p)) & \text{otherwise} \end{cases}$$

Using the above definitions, Tao and Pi [109] find the clusters based on the definitions in the previous section except that their definition of D-reachability is as follows:

p and $q \in \mathcal{D}$ are in each other's k-neighborhood and $\text{NLDF}_k(q) < \tau_{\text{NLDF}}$.

Points outside the clusters are declared as outliers.

7.3 New Algorithms Based on Distance and Cluster Density

Purely rank-based analysis leads to potential incorrect answers when an object is near a dense cluster; this property is the 'cluster density effect'. For instance, two points are of special interest in Fig. 7.5: point 'A' in the neighborhood of a cluster with low density (25 objects) and point 'B' in the neighborhood of a cluster with high density (491 objects).

By visual inspection, it would be argued that the object 'A' is an outlier whereas object 'B' is a possible but not definite outlier. For $k=20$, $O_{20}(A)=25$ because rank of 'A' is 25 from all of its neighbors. On the other hand, the ranks of 'B' with respect to its neighbors are: 2, 8,..., 132, 205, 227; so that $O_{20}(B)$ is 93.1. RBDA concludes that 'B' is more likely to be an outlier than 'A'. This is due to the presence of a large and dense cluster in the neighborhood of 'B'; a point close to a dense cluster is likely to be misidentified as an outlier.

By visual inspection, we intuitively conclude that a point is an outlier if it is 'far away' from the nearest cluster. This implies that the distance of the object (from the cluster) plays an important role; but in RBDA the distance is accounted for indirectly, only through rank. This motivates examining possible improvements that may be obtained by modifying the outlierness measure to give additional weight to the distance of a point from the set containing its neighbors.

Fig. 7.5 An example to illustrate 'Cluster Density Effect' on RBDA; RBDA assigns larger outlierness measure to B

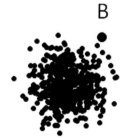

In the first step, the data is clustered using a clustering algorithm, and in the second step, a data object's anomaly score is evaluated based on its distance from the closest cluster. Distance from the centroid of the cluster is, most likely, a poor measure mainly due to the reason that clusters generated by the schema discussed in DBSCAN or the previous section are not necessary spherical; however the point's k-nearest neighbors could be used. A distance could be defined in multiple ways; three distance measures of a point q from the set $\mathcal{N}_k(p)$ are often used:

- $\min_{q \in \mathcal{N}_k(p)} d(p, q)$ (minimal distance)
- $\max_{q \in \mathcal{N}_k(p)} d(p, q)$ (maximal distance)
- $\frac{1}{|\mathcal{N}_k(p)|} \sum_{q \in \mathcal{N}_k(p)} d(p, q)$ (averaged distance)

Note that these distances are not with respect to the entire cluster; only with respect to the point's k-nearest neighbors. Such different distance measures lead to multiple measures of outlierness; the third (averaged distance) measure is used in the following algorithm.

Rank with Averaged Distance Algorithm (RADA) This algorithm adjusts the rank-based outlierness value by the average distance of p from its k-neighbors. The key steps are described below; k, ℓ, m^* are positive integer parameters defined in the previous section:

1. Find the clusters in $\mathscr{D}$ by $\mathcal{NC}(\ell, m^*)$ method.
2. Declare an object o to be a *potential-outlier* if it does not belong to any cluster.
3. Calculate a measure of outlierness:

$$W_k(p) = \mathcal{O}_k(p) \times \frac{\sum_{q \in \mathcal{N}_k(p)} d(q, p)}{|\mathcal{N}_k(p)|} \qquad (7.4)$$

where $\mathcal{O}_k(p)$ is as defined in Eq. (7.1) in the description of RBDA.
4. If p is a potential-outlier and $W_k(p)$ exceeds a threshold, declare p to be an outlier.

For the dataset in Fig. 7.5, $W_{20}(A) = 484.82$ and $W_{20}(B) = 396.19$ implying that A is more likely outlier than B, illustrating that RADA is capable of overcoming the problem observed with RBDA.

Modifications of RBDA We have observed that the size of the neighboring cluster plays an important role when calculating the object's outlierness via RBDA. In RBDA the weight of a cluster C is $|C|$; but a smaller weight would be desirable to reduce the size effect. However, it is not clear how to find the appropriate reduction; weight assignments equal to 1, $\sqrt{|C|}$, and $\log |C|$ are some possibilities resulting in:

1. **ODMR** Suppose that all clusters (including isolated points viewed as clusters of size 1) are assigned weight 1, i.e., all $|C|$ observations of the cluster C are assigned equal weights $= 1/|C|$.

   Then the "modified-rank" of p is defined as $mr_q(p) = $ the sum of weights associated with all observations within the circle of radius $d(q, p)$ centered at q.

Fig. 7.6 Assignment of
weights in different clusters
and modified-rank.
Modified-rank of A, with
respect to B, is $1 + 5 \times \frac{1}{9} + \frac{1}{7}$

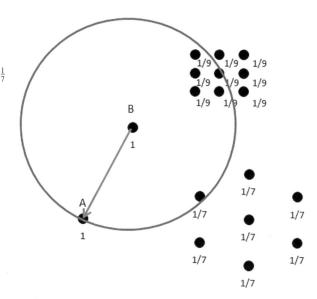

The desired statistic is the sum of "modified-ranks" in $q \in \mathcal{N}_k(p)$, denoted by

$$\Omega = \sum_{q \in \mathcal{N}_k(p)} mr_q(p). \tag{7.5}$$

Figure 7.6 illustrates how modified-rank is calculated.

2. **ODMRS:** If cluster C is assigned a weight $= \sqrt{|C|}$, i.e., each observation of
   the cluster is assigned the weight $= 1/\sqrt{|C|}$, then the "modified-rank" of p is
   obtained by summing these weights associated with all observations within the
   circle of radius $d(q,p)$ centered at q; that is

$$\text{modified-rank of } p \text{ from } q = mr_q^S(p) = \sum_{s \in \{d(q,s) \leq d(q,p)\}} \text{weight}(s).$$

   The associated statistic $\Omega^{(S)} - \sum_{q \in \mathcal{N}_k(p)} mr_q^S(p)$.
3. **ODMRW:** A third alternative, to define modified rank, is as follows. Given a
   cluster C, first we define $p_{k,d} = \sum_{q \in \mathcal{N}_k(p)} d(p,q)$. Then the modified rank of p is
   defined as:

$$mr_q^W(p) = \frac{p_{k,d}}{\sum_{p \in C} p_{k,d}}.$$

   The associated statistic is $\Omega^{(W)} = \sum_{q \in \mathcal{N}_k(p)} mr_q^W(p)$.
4. **ODMRD:** Influenced by the distance consideration of Sect. 7.3, we present
   one more modification of ODMR by using additional distance information.
   ODMRD-outlierness, denoted as $\Omega^{(D)}_k(p)$, is defined as:

$$\Omega^{(D)}{}_k(p) = \sum_{q \in \mathcal{N}_k(p)} mr_q(p) \times d(q, p)$$

This modification has used the modified rank definition of ODMR; alternatives such as ODMRS and ODMRW could also be explored.

Each of the above variations of ODMR can be used to compute outlierness. The algorithm is as follows:

1. Cluster the points in $\mathcal{D}$ using the $\mathcal{N}C(\ell, m^*)$ algorithm;
2. Let the set of potential outliers $P =$ the set of points in $\mathcal{D}$ that do not belong to any cluster;
3. For each point $p \in P$, compute $\Omega(p)$ (as defined in Eq. (7.5) or its variants $\Omega^{(W)}$, $\Omega^{(D)}{}_k(p)$ etc.);
4. Points p with large values of $\Omega(p)$ are declared to be outliers.

When applied to a new problem for which the ground truth is not known, we may examine the distribution of Ω values to determine a threshold and all points whose Ω values exceed this threshold would be considered to be outliers. To find the threshold, one can use the classical ninety fifth percentile of Ω values subject to the condition that there must be a drastic difference between the average of the first 95% values versus the last 5% values (e.g. 4 times more). When there is no such threshold, e.g., if Ω values are uniformly distributed in an interval, the user would have to select a threshold parameter m that indicates how many points are to be considered outliers. Two algorithms could be compared by examining the sets of m points considered to have the highest outlierness values (Ω), and evaluating in each case how many of these are true outliers (if ground truth is known).

7.4 Results

We compare RBDA and its extensions with algorithms presented in Chap. 6.

7.4.1 RBDA Versus the Kernel Based Density Estimation Algorithm

Recall the anomaly detection algorithm described in Chap. 6. This algorithm computes the density at a point using the non-parametric approach with Gaussian kernel. RBDA was compared with this algorithm for datasets described in the Appendix. It was observed that RBDA performs better than the kernel based approach for the synthetic dataset 1, Iris-5 and Wisconsin-10. On the other hand the kernel based algorithm performs better than RBDA when the window is large ($\sigma = 5$) for Iona-10 and Iona-3 datasets; for the Iris-3 dataset, both performed equally well.

7.4.2   Comparison of RBDA and Its Extensions with LOF, COF, and INFLO

The performance of RADA, ODMR, ODMRS, ODMRW, ODMRD, RBDA, DBCOD, LOF, COF and INFLO were compare for several datasets described in the Appendix.

- For the synthetic dataset 2, we compared performances for $k = 25, 35$, and 50 and $m = 6, 10, 15, 20, 30$. For $k = 25$ and 35 the performance of RADA, ODMR, ODMRS, ODMRW, and ODMRD were identical to the performance of RBDA. For $k = 50$ the performance of RADA, ODMR, ODMRS, ODMRW, and ODMRD were identical and slightly better than of RBDA. Comparisons of the remaining algorithms are presented in Table 7.1.

Table 7.1 Performance of each algorithm for synthetic dataset 2. The largest values (best performances) are shown in red color

k=25	RBDA			DBCOD			LOF			COF			INFLO		
m	m_t	Re	RP	m_t	Re	RP	m_t	Re	RP	m_t	Re	RP	m_t	Re	RP
6	6	1	1	3	0.5	0.857	4	0.667	1	3	0.5	0.857	3	0.5	1
10	6	1	1	4	0.667	0.667	5	0.833	0.882	3	0.5	0.857	4	0.667	0.714
15	6	1	1	5	0.833	0.577	6	1	0.656	5	0.833	0.484	5	0.833	0.6
20	6	1	1	6	1	0.5	6	1	0.656	5	0.833	0.484	5	0.833	0.6
30	6	1	1	6	1	0.5	6	1	0.656	6	1	0.375	6	1	0.438

k=35	RBDA			DBCOD			LOF			COF			INFLO		
m	m_t	Re	RP	m_t	Re	RP	m_t	Re	RP	m_t	Re	RP	m_t	Re	RP
6	5	0.833	1	3	0.5	0.857	2	0.333	0.429	0	0	0	2	0.333	0.429
10	5	0.833	1	5	0.833	0.577	3	0.5	0.375	2	0.333	0.158	3	0.5	0.375
15	6	1	0.808	5	0.833	0.577	5	0.833	0.357	5	0.833	0.259	5	0.833	0.375
20	6	1	0.808	6	1	0.457	6	1	0.362	5	0.833	0.259	5	0.833	0.375
30	6	1	0.808	6	1	0.457	6	1	0.362	6	1	0.256	6	1	0.344

k=50	RBDA			DBCOD			LOF			COF			INFLO		
m	m_t	Re	RP	m_t	Re	RP	m_t	Re	RP	m_t	Re	RP	m_t	Re	RP
6	5	0.833	1	3	0.5	0.857	1	0.167	0.5	0	0	0	1	0.167	0.5
10	5	0.833	1	3	0.5	0.857	2	0.333	0.25	1	0.167	0.1	2	0.333	0.25
15	5	0.833	1	5	0.833	0.484	4	0.667	0.278	1	0.167	0.1	5	0.833	0.3
20	5	0.833	1	5	0.833	0.484	6	1	0.3	2	0.333	0.1	5	0.833	0.3
30	6	1	0.583	6	1	0.396	6	1	0.3	6	1	0.175	6	1	0.292

- **Comparisons for the Iris Dataset**

  - For the dataset with rare class, results show that ODMRW achieves the best performance when k is 5; LOF is the best when k is 10, while ODMRW is the second best. For $k = 10$, RADA, ODMRW, RBDA, LOF, and INFLO perform best.
  - For the dataset with planted outliers; ODMRW achieves the best performance for all values of k. RADA, ODMR, ODMRS, ODMRD and RBDA all have similar performances. LOF, COF and INFLO do not work well for this data set since they hardly detect any outlier when $m \leq 40$.

- **Comparisons for the Ionosphere Dataset**

  - For the dataset with rare class, results (with respect to metrics m_t and Recall) show that the algorithms RBDA, ODMR, ODMRD, and RADA perform equally well whereas DBCOD does not perform well for $k \leq 30$.
  - For the dataset with planted outliers, ODMRW performs better than the other algorithms for all values of k. DBCOD, LOF, COF and INFLO do not work well for this data set since they are not able to detect more than 1 (out of 3) outlier when $m \leq 30$.

- **Comparisons for the Wisconsin Dataset**

  - Results for the dataset with rare class show that no single algorithm dominates, but Rank-based algorithms tend to perform better than LOF, COF, INFLO and DBCOD. For $k = 7$ RBDA and RADA both work better than other algorithms; COF performs the best when $k = 11$; and DBCOD shows the best performance when $k = 22$. In general, no single algorithm can achieve the best performance for all values of k. Most of the algorithms perform equally well and are able to detect all 10 outliers when m is 40.
  - Results for the dataset with inserted anomalies show that RADA, ODMR, ODMRS, ODMRW, ODMRD and RBDA achieve the best performances for all values of k.

- **Overall Performance** Table 7.2 summarizes the performance of all algorithms over all values of k, and is obtained by using *normalized* RankPower. RankPowers are normalized to fall in the scale [0, 100], where 0 means that the RankPower of the algorithm is least, and 100 corresponds to the largest RankPower. Values of k were chosen between 5% to 10% of the size of datasets.

Although better performance is indicated by larger values of all three measures of performance, m_t, recall, and RankPower, we observe that RankPower is more discriminatory than the other two metrics. Using this metric, for the datasets mentioned above, the relative behavior of algorithms can be summarized as:

$$\text{ODMRW} \geq \text{ODMRD} \geq \text{ODMR} \geq \text{RADA}$$

$$\geq \text{ODMRS} \geq \text{RBDA} \geq \text{LOF} \geq \text{DBCOD} \geq \text{INFLO} \geq \text{COF}$$

where by "$\geq$" we indicate a better performance.

Table 7.2 Summary of LOF, COF, INFLO, DBCOD, RBDA, RADA, ODMR, ODMRS, ODMRW, and ODMRD for all experiments. Numbers represent the average performance rank of the algorithms; a larger value implies better performance. Data set with 'r' in the parentheses represents the data set with rare class. And data set with 'o' in the parentheses represents the data set with planted outliers

Dataset	LOF	COF	INFLO	DBCOD	RBDA	RADA	ODMR	ODMRS	ODMRW	ODMRD
Synthetic	44	27	36	45	80	100	100	100	100	100
Iris(r)	82	17	59	26	74	76	90	77	98	91
Inosphere(r)	54	57	58	26	92	97	92	92	100	98
Wisconsin(r)	62	86	61	92	93	95	93	92	86	92
Iris(o)	100	67	100	100	100	100	100	100	100	100
Inosphere(o)	49	48	53	34	92	96	92	92	100	96
Wisconsin(o)	6	20	12	47	100	100	100	100	100	100
Summary	57	46	54	53	90	95	95	93	98	97

Table 7.3 KDD 99, $k = 55$. For each value of m the best results are in red color. Results for COF and INFLO were generally poorer than for LOF and are not shown below

	LOF			RBDA			RADA			ODMR			ODMRW		
m	m_t	Re	Rp	m_t	Re	Rp	m_t	Re	Rp	m_t	Re	Rp	m_t	Re	Rp
12	4	0.33	0.04	6	0.50	0.13	6	0.50	0.14	6	0.50	0.14	6	0.50	0.14
20	6	0.50	0.07	8	0.67	0.22	10	0.83	0.37	10	0.83	0.37	10	0.83	0.37
30	8	0.67	0.13	5	0.42	0.03	12	1.00	0.53	12	1.00	0.52	12	1.00	0.52
40	8	0.67	0.13	6	0.50	0.04	12	1.00	0.53	12	1.00	0.52	12	1.00	0.52
50	11	0.92	0.23	8	0.67	0.07	12	1.00	0.53	12	1.00	0.52	12	1.00	0.52
60	11	0.92	0.23	9	0.75	0.09	12	1.00	0.53	12	1.00	0.52	12	1.00	0.52
80	12	1.00	0.27	11	0.92	0.13	12	1.00	0.53	12	1.00	0.52	12	1.00	0.52

7.4.3 Comparison for KDD99 and Packed Executables Datasets

Aforementioned algorithms were compared for KDD99 and Packed Executable datasets. Representative tables and summary are provided below.

- For the KDD99 dataset, the performance was measured for $k = 15, 35, 55, 105, 155$, and 205, and $m = 12, 20, 30, 40, 50, 60$ and 80. Detailed performance is presented in Table 7.3 for $k = 55$ only.

  It can be seen that for $k = 55$, the best performance is given by RBDA. Performances of ODMR, OGMRW and RBDA are also pretty good, whereas LOF, INFLO and COF have poor performance, especially when measured in terms of RankPower. Similar observations can be made for other values of k. Overall performance of RADA is the best.
- The following Table 7.4 gives results for the Packed Executable Dataset, for $k = 55$. Other values of $k = 15, 35, 105, 155$, and 205, were also used.

Table 7.4 Packed Executable Dataset $k = 55$. For each value of m the best results are in red color. Results for COF and INFLO were generally poorer than for LOF and are not shown below

	RBDA			INFLO			RADA			ODMR			ODMRW		
m	m_t	Re	Rp	m_t	Re	Rp	m_t	Re	Rp	m_t	Re	Rp	m_t	Re	Rp
8	3	0.38	0.07	0	0.00	0.00	4	0.50	0.17	4	0.50	0.10	4	0.50	0.10
10	4	0.50	0.11	0	0.00	0.00	4	0.50	0.17	4	0.50	0.10	4	0.50	0.10
20	6	0.75	0.23	0	0.00	0.00	8	1.00	0.61	6	0.75	0.22	6	0.75	0.22
30	7	0.88	0.31	0	0.00	0.00	8	1.00	0.61	7	0.88	0.29	6	0.75	0.22
40	8	1.00	0.40	0	0.00	0.00	8	1.00	0.61	8	1.00	0.37	8	1.00	0.38
50	8	1.00	0.40	1	0.13	0.00	8	1.00	0.61	8	1.00	0.37	8	1.00	0.38
60	8	1.00	0.40	3	0.38	0.07	8	1.00	0.61	8	1.00	0.37	8	1.00	0.38

From the above table, it is easy to conclude that the performance of RADA is the best. However, from other tables (not presented here) it was observed that for $k = 15$, the best performance is given by ODMRW, followed by ODMR, RADA, and RBDA. For $k = 35, 55$, and 105, performance of RADA is the best followed by ODMR, ODMRW, and RBDA. For these values of k COF, INFLO and LOF has extremely poor performance. However, when $k = 155$ and 255, LOF has the best performance; RADA is close behind it. In summary, it is observed that RADA is a good algorithm, reliable in most cases, but the value of k matters. It would be useful to determine appropriate value of k for a given dataset.

7.5 Conclusions

The performance of an outlier detection algorithm based on rank alone is highly influenced by cluster density variations. Furthermore, by definition, ranks use the relative distances and ignore the 'true' distances between the observations. This motivates development of new outlier detection algorithms that utilize rank as well as distance information.

Extensive evaluations on synthetic and real datasets have demonstrated that the overall performance of each of the new algorithms (ODMR and other variants) is significantly better than previously known algorithms. Among the new variants of algorithms, ODMRW performed best, perhaps due to the greater weightage placed on distance.

Chapter 8
Ensemble Methods

In the previous chapters, we have described various anomaly detection algorithms, whose relative performance varies with the dataset and the application being considered. It may be impossible to find one algorithm that outperforms all others, but *ensemble methods* that combine the results of multiple algorithms often provide the best results. A particular anomaly detection algorithm may be well-suited to the properties of one dataset and be successful in detecting anomalous observations of the particular application domain, but may fail to work with other datasets whose characteristics do not agree with the first dataset. The impact of such mismatch between an algorithm and an application can be alleviated by using ensemble methods where a variety of algorithms are pooled before a final decision is made.

This chapter discusses ensemble methods to improve the performance of anomaly detection algorithms, and successfully address the high false positive rates of individual algorithms, referred to as *weak learners*. Ensemble methods for anomaly detection have been categorized into two groups [4]: the independent ensemble and the sequential ensemble approaches. In the independent ensemble approach, discussed in Sect. 8.1, the results from executions of different algorithms are combined in order to obtain a more robust model. In sequential ensemble approaches, summarized in Sect. 8.2, a set of algorithms is applied sequentially. Section 8.3 discusses adaptive sampling methods based on the *Adaboost* algorithm and on the *active learning* approach. A weighted adaptive sampling algorithm is then described, in Sect. 8.4. Section 8.5 presents concluding remarks.

8.1 Independent Ensemble Methods

In the *Independent Ensemble* methods, each algorithm provides an anomaly score to objects in $\mathscr{D}$; objects that receive higher scores are considered more anomalous. The ranges and distributions of scores may be substantially different for different

© Springer International Publishing AG 2017

K.G. Mehrotra et al., *Anomaly Detection Principles and Algorithms*, Terrorism, Security, and Computation, https://doi.org/10.1007/978-3-319-67526-8_8

algorithms; normalization of individual scores is hence necessary before combining scores. Normalization may be performed using a linear mapping that assigns a normalized score of 0 to the least anomalous data point, and a normalized score of 1 to the most anomalous data point (as per each specific algorithm). We use the notation that $\alpha_i(p)$ is the normalized anomaly score of $p \in \mathcal{D}$, according to algorithm i.

In some algorithms, anomaly scores are sorted in decreasing order and assigned *ranks,* and these ranks are used in ensemble methods, instead of using normalized anomaly scores. We follow the convention that the object that receives the highest score is ranked 1, the second most anomalous object is ranked 2, and so on. We use the notation that $r_i(p)$ is the rank of $p \in \mathcal{D}$, according to algorithm i.

The approaches that can be used to combine the results of multiple algorithms can focus either on the raw anomaly scores (α_i), or on the ranks (r_i). Most combination methods seek a consensus among algorithms, by computing the median or mean of the anomaly scores or ranks; an alternative approach considers a data point to be highly anomalous if at least one algorithm evaluates it to be highly anomalous–this is implemented by computing the maximum of the anomaly scores, or the minimum of the ranks.

Averaging The averaged normalized score of p over m individual algorithms is obtained as follows:

$$\alpha(p) = \frac{1}{m} \sum_{i=1}^{m} \alpha_i(p).$$

Final ranking is based on this averaged score, and the data point with the highest α value is ranked 1 (i.e., is most anomalous). It is possible instead to perform averaging directly on the ranks, but this can result in the loss of useful information, e.g., whether the top two ranked individuals obtained using an algorithm i differ substantially or minimally in α_i values.

Example 8.1 For a dataset with three points, upon applying three individual anomaly detection algorithms, let

$$\alpha_1(p_1) = 1.0, \alpha_1(p_2) = 0.9, \alpha_1(p_3) = 0.0,$$

$$\alpha_2(p_1) = 1.0, \alpha_2(p_2) = 0.8, \alpha_2(p_3) = 0.0,$$

and

$$\alpha_3(p_1) = 0.1, \alpha_3(p_2) = 1.0, \alpha_3(p_3) = 0.0.$$

Then the averaged anomaly scores are $\alpha(p_1) = 0.7$, $\alpha(p_2) = 0.9$, and $\alpha(p_3) = 0.0$, suggesting that p_2 is more anomalous than p_1, whereas averaging the ranks yields an average rank of 1.3 for p_1 and 1.7 for p_2, suggesting that p_1 is more anomalous than p_2. The former inference may be more reasonable since algorithms 1 and 2 do not substantially distinguish between the relative anomalousness of p_1 and p_2 (since they have very similar anomaly scores), whereas algorithm 3 does consider them to have substantially different anomaly scores. This may happen because the first two algorithms use almost identical approaches to compute anomaly scores, whereas the third algorithm behaves quite differently, and the strong opinions of the latter should not be submerged by the weak opinions of the first two.

The Min-Rank Method   Let the anomalous rank of p, assigned by algorithm i be given by $r_i(p)$. As mentioned earlier, $r_i(p) < r_i(q)$ implies that $\alpha_i(p) > \alpha_i(q)$, hence p is more anomalous than q according to algorithm i. The *Min-rank* method assigns the smallest possible rank to each data point, implicitly emphasizing points that may be considered anomalous according to any single criterion or algorithm, even if other algorithms do not consider the point to be anomalous. If m algorithms are being combined, then

$$\mathrm{rank}(p) = \min_{1 \le i \le m} r_i(p).$$

In other words, if the object p is found to be most anomalous by at least one algorithm then the Min-rank method also declares it to be the most anomalous object. If all m algorithms give substantially different results, several different points may have the same combined (minimum) rank. Maximization of α_i (instead of r_i) values should give similar results but is not recommended since each algorithm may compute anomaly scores in a completely different way, and direct comparison of anomaly scores of different algorithms may not make sense: if $\alpha_1(p_1) > \alpha_2(p_1)$ and $\alpha_2(p_2) > \alpha_1(p_2)$, then we cannot infer from the maximum α_i values whether p_1 is more anomalous than p_2.

Example 8.2   Let α_i values for three points be defined for three algorithms as in the previous example, resulting in the following ranks:

$$r_1(p_1) = 1, r_1(p_2) = 2, r_1(p_3) = 3;$$

$$r_2(p_1) = 1, r_2(p_2) = 2, r_2(p_3) = 3;$$

$$r_3(p_1) = 2, r_3(p_2) = 1, r_3(p_3) = 3.$$

Then the minimum rank values combined from the three algorithms are:

$$rank(p_1) = 1, rank(p_2) = 1, rank(p_3) = 3.$$

Both p_1 and p_2 are then considered equally anomalous, but more anomalous than p_3. Note that averaging the ranks would have given a different result: the average rank of p_1 is greater than that of p_2. Maximizing the α_i values gives 1.0, 1.0, and 0.0, for the three points, respectively, which is consistent with minimizing the ranks in this example.

Majority Voting Another possible approach is to consider a point p to be more anomalous than q if more of the m algorithms assign higher scores to p than to q. First we define the comparison indicator function *comp* as follows, where $\epsilon \geq 0$ is a small threshold used to determine whether an algorithm considers two points to be equally anomalous:

$$comp_i(p, q) = \begin{cases} 1 & \text{if } \alpha_i(p) > \alpha_i(q) + \epsilon, \\ -1 & \text{if } \alpha_i(p) < \alpha_i(q) - \epsilon, \\ 0 & \text{if } |\alpha_i(p) - \alpha_i(q)| \leq \epsilon. \end{cases}$$

Then,

$$comp(p, q) = \sum_{i=1}^{m} comp_i(p, q)$$

roughly indicates the degree of consensus between the m algorithms regarding whether p is more anomalous than q. By definition, note that $comp_i(q, p) = -comp_i(p, q)$, hence $comp(q, p) = -comp(p, q)$. This pairwise comparison measure can be used to construct a composite rank or anomaly score, e.g., defining $\alpha(p)$ to be $\sum_{p_j \in \mathcal{D}} comp(p, p_j)$.

Example 8.3 Let α_i values for three points be defined for three algorithms as in the previous example, and let the threshold $\epsilon = 0.1$. Then

$$comp_1(p_1, p_2) = 0, comp_1(p_1, p_3) = 1, comp_1(p_2, p_3) = 0;$$
$$comp_2(p_1, p_2) = comp_2(p_1, p_3) = comp_2(p_2, p_3) = 1;$$
$$comp_3(p_1, p_2) = -1, comp_3(p_1, p_3) = 0, comp_3(p_2, p_3) = 1.$$

Then the combined comparison values are $comp(p_1, p_2) = 0$, $comp(p_1, p_3) = 2$, and $comp(p_2, p_3) = 2$, yielding the composite anomaly scores $\alpha(p_1) = 2$, $\alpha(p_2) = 2$, and $\alpha(p_3) = -4$. The final result again indicates that p_1 and p_2 are equally anomalous, but more than p_3. Note that the results depend on ϵ; for example, if ϵ had been chosen to be 0.05 in the above example, then p_2 would be considered to be more anomalous than p_1.

Majority voting has been used in many other contexts, including voting in the political context, and many results have been developed addressing its usefulness. Condorcet's *Jury Theorem* [28] establishes a mathematical result applicable to ensemble methods, and states that if a majority vote is taken using the opinions of weak learners who have an independent probability > 0.5 of voting for the correct decision, then adding more such voters improves the probability of the majority decision being correct. Both the qualifying criteria are important: the voters must be independent, and each voter must be more likely to vote correctly than incorrectly. Unfortunately, many practical applications violate the first criterion, and enlarging the jury may decrease its collective competence when votes are correlated [66]. A stronger result shows that if the number of voters is sufficiently large and the average of their individual propensities to select the better of two policy proposals is a little above random chance, and if each person votes his or her own best judgment (rather than in alliance with a block or faction), then the majority is extremely likely to select the better alternative [56]; this result is particularly useful for ensemble algorithms for anomaly detection, since each algorithm applies its own procedure (i.e., no voting in a block) even though some algorithms produce results that are correlated since they examine the same data and apply similar procedures.

Sampling Empirical results have shown that these methods provide improvements over the results of single anomaly detection algorithms. However, some of the above approaches require significantly high computational effort; for instance, if a dataset contains a thousand points, then millions of point-point comparisons would be needed for the majority voting approach. Sampling can be used to identify promising candidates for the most anomalous data points using any of the above approaches, and exhaustive comparisons can be restricted to such promising candidates. For example, we may repeatedly draw a relatively small random sample of fixed size, $S \subset \mathcal{D}$ (with replacement), and find the most anomalous element of S by the min-rank approach, adding it to the *Candidates* set. All the members of *Candidates* may then be exhaustively compared with each other, perhaps using the majority voting approach. Alternatively, an anomaly score for each data point may be computed as the average of the anomaly score with respect to multiple small samples (subsets of $\mathcal{D}$) [126].

8.2 Sequential Application of Algorithms

Even the sampling approach (described above) requires that each algorithm be applied to the entire dataset. Greater computational efficiency would be obtained if promising solutions are first obtained by a subset of algorithms, and the remaining algorithms restrict their anomaly score computations to these promising solutions. In the sequential approach, we begin with the entire dataset $\mathcal{D}$, and successively apply one algorithm after another, refining the subset of promising solutions at each stage by eliminating the least anomalous candidates from the previous step. Using substantially diverse algorithms is very important when combining different anomaly detection algorithms [4, 125]. The extent to which two algorithms produce similar results can be evaluated by computing the Pearson correlation coefficient between each pair of algorithms using the associated score vectors. Diversity is maximized by selecting the pair of algorithms with the smallest average correlation value. Zhao et al. [122] applied this approach to several algorithms discussed in earlier chapters, and observed that the smallest correlation corresponds to COF and RADA. In other words, COF followed by RADA (or in reverse order) should be successful in anomaly detection. The largest average correlations are observed between the three algorithms of rank-family, i.e., between RBDA, RADA, and ODMR; implying that selecting two algorithms among these will not perform as well due to low diversity.

As in the previous section, ranks for all data points are calculated using the anomaly scores. In the *Sequential-1* method (cf. Algorithm "Sequential-1 algorithm"), one anomaly detection algorithm (such as COF) is first applied on the whole dataset to obtain the ranks of all observations. Next the dataset $\mathcal{D}_\alpha$ is obtained by retaining the $\alpha \times |\mathcal{D}|$ elements of $\mathcal{D}$ with the highest anomaly scores using algorithm 1, suspected to contain all anomalies. The second anomaly detection algorithm calculates the anomaly scores of all objects in the dataset with reference to $\mathcal{D}_\alpha$. For instance, if $\alpha = 0.1$, then the second algorithm is applied only to 10% of the elements in $\mathcal{D}$. This process can be repeated using additional individual anomaly detection algorithms. Computational effort decreases as α decreases, with an increased risk of losing data points that are anomalous only from the perspective of the algorithm(s) applied later.

Algorithm Sequential-1 algorithm

Input: dataset $\mathcal{D}$, detection algorithms A and B, fraction $0 \leq \alpha \leq 1$
Output: a set of k most anomalous objects in $\mathcal{D}$

1. ScoreList$_A$ $(\mathcal{D})$ = List of anomaly scores obtained using algorithm A on $\mathcal{D}$;
2. RankList$_A$ $(\mathcal{D})$ = List of objects in $\mathcal{D}$ sorted by the order of decreasing anomaly scores (from ScoreList);
3. $\mathcal{D}_\alpha$ = List of the first $\alpha \times |\mathcal{D}|$ objects in RankList$_A$ $(\mathcal{D})$;
4. ScoreList$_B$ $(\mathcal{D}_\alpha)$ = List of anomaly scores obtained using algorithm B on $\mathcal{D}_\alpha$;
5. RankList$_B$ $(\mathcal{D}_\alpha)$ = List of objects in $\mathcal{D}_\alpha$ sorted by the order of decreasing anomaly scores (from ScoreList);
6. **Return** the list of the first k elements in RankList$_B$ $(\mathcal{D}_\alpha)$.

This approach can be used along with the sampling approach mentioned earlier. Zhao et al. [122] extend the single algorithm sampling approach of Zimek et al. [126], using the second anomaly detection algorithm with the averages of anomaly scores obtained using subsamples drawn from $\mathcal{D}_\alpha$ (using the first algorithm). Empirical results showed that best results were obtained in most cases with this approach using either LOF or COF followed by RADA.

8.3 Ensemble Anomaly Detection with Adaptive Sampling

In the previous sections, we considered either combining the outputs of multiple algorithms, or applying one algorithm to the results obtained by another. This section extends the first approach, repeatedly examining targeted observations to refine decisions.

Averaging the outputs of different weak learners gives them equal weightage, whereas each weak learner may be shown to work well on a specific subset of $\mathcal{D}$. This issue can be addressed using *weighted majority voting* in which the weight of a learner is assigned by measuring the quality of its results. For any object $o \in \mathcal{D}$, if a learner's output at iteration $t \in \{1, 2, \ldots, T\}$ is $\ell_t(o)$, the final output is denoted by:

$$\mathcal{L}(o) = \sum_{t=1}^{T} \beta_t \ell_t(o)$$

where β_t indicates how important the learner ℓ_t is in the final combination. For example, in the context of a classification problem, one can use the error on the training data as an indication of how well a learner is expected to perform.

A few ensemble methods use repeated samples for unsupervised anomaly detection [3, 125]. We now discuss ensemble approaches based on the *Adaboost* approach [122].

8.3.1 AdaBoost

AdaBoost is an ensemble-based supervised classification algorithm [44] which has gained considerable popularity for its elegance and performance. The algorithm iterates through T rounds of *weak learners* applied to the training dataset. In each round, Adaboost focuses on the 'hard' examples from the previous round, then combines their output by weighted majority voting. AdaBoost calls a "booster" in each round to draw a subsample $\mathcal{D}_t$ from $\mathcal{D}$ with a set of sampling probabilities, initially uniform. A *weak learner* or *base method* h_t is trained over $\mathcal{D}_t$, then the sampling probabilities are increased for incorrectly classified examples. After T rounds, each *weak learner* ℓ_t is assigned with a weight β_t which is lower if the corresponding error for that learner is higher. The final decision output is made by applying weighted majority voting over all ℓ_t for $t = 1, 2, \ldots, T$. The training error will be substantially reduced if all the *weak learners* are better than random guessing.

An optimal weight choice criterion, assuming decisions are made independently of each other, states that the overall probability of the ensemble making the correct decision by majority rule is obtained by assigning weights to different decision-makers that are proportional to $w_i = log(p_i/(1 - p_i))$ where p_i is the probability of the ith decision maker making the correct choice [53]. A corollary states that if p_i is unknown, we can approximate the optimal weight choices by w_i proportional ($\rho_i -$ 0.5) where ρ_i measures how often the ith decision-maker agrees with the majority.

The method proposed by Abe et al. [1] converts the anomaly detection problem to an AdaBoost-solvable classification problem by drawing some number of points from an underlying distribution, and marking them as anomalies, whereas all the points in the original dataset are considered to be inliers. Their resampling method is based on *minimum margin active learning* by iteratively assigning lower sample rejection probabilities to points with low margin values, because they have less consistency. The weights for each classifier are adjusted by the classification error rate in each subsample.

Zhao et al. [122] propose a different approach that adapts AdaBoost's approach to the unsupervised anomaly detection problem. This algorithm, discussed in greater detail in Sect. 8.4, uses the score output of the *base algorithm* to determine the 'hard' examples, and iteratively re-sample more points from such examples in a complete unsupervised context. First, we discuss the adaptive sampling approach.

8.3.2 Adaptive Sampling

Adaptive learning, also known as *Active Learning*, refers to a procedure where intervention or human feedback is applied to achieve superior performance from a given dataset. Consider the problem of two-class classification: a classifier is designed to associate each observation with one of the two classes. In addition, assume that there is only one feature, described by the Gaussian distribution, and the classes differ only in the mean but the variances are equal. In this simplest classification problem, if the training set consists of two simple random samples from the two Gaussian distributions, then the classifier is likely to perform poorly in comparison with the classifier built using most of the observations near the boundary. In essence, this is the key difference between Fisher's linear discriminant and the Support Vector Machine (SVM) approach that emphasizes observations near the boundary of the two distributions.

In some active learning frameworks, learning proceeds in rounds; each round starts with a small number of 'labeled' examples. Then the learner models the combined data, i.e., the new labeled data as well as the whole or subset of the original data. The model is used to identify a few data points such that obtaining labels for them would help to improve the model. These are shown to the teacher for labeling, and the cycle repeats.

8.3.3 Minimum Margin Approach

The ensemble-based minimum margin approach [1] transforms the unsupervised anomaly detection problem to a supervised classification problem solved using any classification algorithm A. All the given observations, $\mathscr{D}$, are identified with Class 0, and Class 1 consists of a *synthetic sample* containing an equal number of observations generated from a uniform distribution over a range that includes the entire range of values of data attributes in $\mathscr{D}$. The classification algorithm A is iteratively applied with different samples, and the ith classifier C_i is just the application of A to the ith sample $S_i \subset \mathscr{D}$ whose elements are chosen with higher probability when there is less consensus between prior classifiers $C_0, \ldots, C_{i-1}$.

Thus, the algorithm successively focuses on the decision boundary between the obvious normal examples and the clear outliers.

For each data point x, if $C_j(x) = b$ indicates that x is placed in class b by classifier C_j, the minimum margin algorithm computes the margin value $M(x)$ based on the previous $C_0, C_1, \ldots, C_{i-1}$ classifiers as follows:

$$M(x) = \left| \sum_{j=0}^{i-1} \left(\text{Prob}(C_j(x) = 1) - \text{Prob}(C_j(x) = 0) \right) \right|.$$

Note that $M(x)$ is high when the classifiers $C_0, C_1, \ldots, C_{i-1}$ unanimously consider x to be either an outlier or an inlier, so not much is to be gained by applying the next classifier C_i to x; on the other hand, $M(x) \approx 0$ when the classifiers are equally split in their opinions, so the next classifier C_i should carefully consider x with a high probability. The actual probability of selecting observation x into sample S_i is evaluated using the area under the right-hand tail of a Gaussian curve

$$\Phi(\mu, \sigma, y) = \int_y^\infty \frac{1}{\sqrt{2\pi}\sigma} e^{\frac{-1}{2}\left(\frac{x-\mu}{\sigma}\right)^2} dx$$

where $\mu = \frac{i}{2}$, $\sigma = \frac{\sqrt{i}}{2}$, and $y = \frac{i+M(x)}{2}$. This probability is largest when $M(x) = 0$ (when y is smallest), and the probability is smallest (but non-zero) when $M(x) \approx i$, i.e., when $C_0, C_1, \ldots, C_{i-1}$ are unanimous. Thus, successive classifiers pay greater attention to the points on which previous classifiers have no consensus.

The final decision regarding whether a point x is anomalous is based on a weighted average of the opinions of various classifiers $C_0, \ldots, C_t$, and the weight of each classifier C_i, which is obtained as a function of the error rate of C_i on the sample S_i, e.g., $w_i = log(1\text{-} (\text{error rate})/(\text{error rate}))$, following the Adaboost approach.

8.4 Weighted Adaptive Sampling

The minimum margin approach discussed in the previous section computes a weighted average of the decisions of different classifiers C_i obtained by applying the same classification algorithm A to selected samples of $\mathcal{D}$. This section describes an analogous approach, in which the results of different anomaly detection algorithms are combined, using weighted adaptive sampling that uses the anomaly scores obtained with different algorithms [123].

Adaptive sampling emphasizes examples that are "hard to identify", i.e., near the boundary of inliers and outliers, as seen in Fig. 8.2. To extract such examples from the dataset, we need to answer two questions:

1. How can we determine whether an object can be easily classified as an outlier?
2. How can we combine the outputs from different iterations?

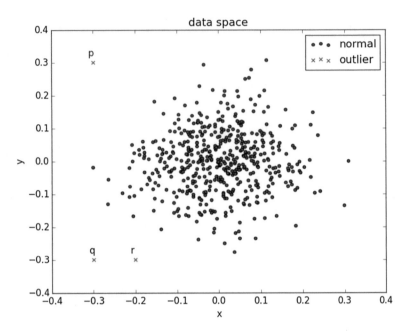

Fig. 8.1 An example with three anomalous data points p, q, and r added to synthetic data generated randomly

The approach described in this section uses anomaly score outputs from the base algorithms to determine the set of boundary points, defined below, which are re-sampled with higher probability using a *kNN* approach.

Decision Boundary Points

A simple example is illustrated in Fig. 8.1, containing three artificially introduced anomalous data points: p, q, and r. The LOF anomaly detection algorithm (with $k = 3$) yields anomaly scores for each object, normalized to the $[0,1]$ range. The mapping of data points into the score space is shown in Fig. 8.2.

The object p on the upper left corner is the most obvious anomaly among all of the three anomalies, and was assigned the highest anomaly score as shown in Fig. 8.2; this illustrates that the detection algorithm performs very well when detecting the easy-to-detect anomalies. But the scores for the other two points are relatively low, and are almost the same as those of some non-anomalous data points; these are considered to be on the boundary. The score space can be used to distinguish potential anomalies and inliers, all objects with an anomaly score greater than θ (a predetermined threshold) can be identified as anomalous and all other objects are marked as inliers; θ determines the decision boundary. In the example of Fig. 8.2, if we choose $\theta = 0.3$, then two real anomalies p,q are identified, but five false positives are introduced, and we have one false negative r (undetected anomaly).

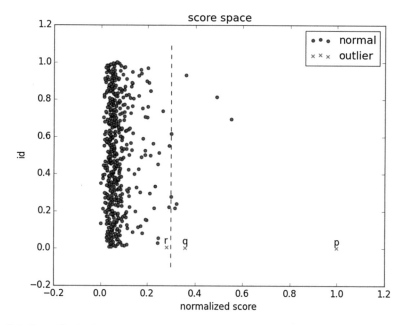

Fig. 8.2 Normalized LOF scores for data in Fig. 8.1; the y-axis represents the data points

It is unclear whether points near the threshold boundary are true anomalies, since the threshold is arbitrarily chosen; a slightly larger or smaller value of θ may yield different results. This motivates another approach that applies higher sampling probability weights to points that are near a decision boundary [123], described below, justified with an analogy to anomaly detection algorithms based on computing the local density near each point.

Sampling Weights Adjustment Consider normalized scores $\alpha(x) \in [0, 1]$ for each data point x. Note that $\alpha(x) \approx 0$ identifies a clear inlier, $\alpha(x) \approx 1$ identifies a definite anomaly, and $\alpha(x) \approx \theta$ implies that x is more likely to be a boundary point, where $\theta \in [0, 1]$ is an appropriately chosen threshold.

Most false positives and false negatives are expected to occur at boundary points, with existing algorithms. Hence a new weight $WB(x)$ is assigned to each data point x, increasing with the likelihood that it is a boundary point, as follows:

$$WB(x) = \begin{cases} \frac{\alpha(x)}{\theta} & \text{if } \alpha(x) < \theta \\ \frac{1-\alpha(x)}{1-\theta} & \text{if } \alpha(x) \geq \theta. \end{cases} \tag{8.1}$$

For example, if $\theta = 0.75$ and $\alpha(x_1) = 0.8$, then $WB(x_1) = (1 - 0.8)/(1 - 0.75) = 0.8$, a high value indicating that x_1 needs extra attention, whereas if $\theta = 0.75$ and $\alpha(x_2) = 0.99$, then $WB(x_2) = (1 - 0.99)/(1 - 0.75) = 0.04$, hence x_2 is a definite outlier that does not need further attention in refining the learning algorithm parameters.

Such a weight assignment increases sampling probability for boundary points and decreases the sampling probability for points that are easy to detect (clear inliers and anomalies). Instead of iteratively refining the boundaries as in classification or clustering problems, this approach iteratively resamples the points near the boundaries. This helps address some false negatives and positives caused by low data density near the boundary. As is the case with many other learning algorithms, focusing attention on boundary points improves the performance, refining the boundary, instead of wasting computational effort on obvious inliers and outliers.

Final Outputs Combination The absence of labeled training data makes it difficult to measure the performance quality of a clustering or anomaly detection algorithm in an ensemble approach. If the outputs from all iterations are independent, we can use the correlation between them as an indication of how well one individual learner performs. However, in the adaptive learning process, the input of each learner is dependent on the previous one, and this makes the process of selection of weights (for each learner) even harder. The problem of measuring how well a learner performs in an iteration is essentially the question of how many 'real' anomalies were captured in that iteration. Objects with higher scores are more anomalous than the ones with lower scores. But a more pertinent question is that of determining the size of the gap between scores of anomalous vs. non-anomalous objects. Histograms in Fig. 8.3 show normalized scores for two datasets: (1) with no anomaly and (2) for the same dataset with 3 anomalies inserted. We observe that the anomalies get larger scores, and also the gap in scores between the inliers and anomalies increases, in Fig. 8.3a, the 'gap' is from 0.7 to 0.8 while in Fig. 8.3b, the 'gap' is from 0.6 to 0.9.

These observation suggest three alternative weight assignments for β_t:

a) The simplest approach is to take the arithmetic average for all values of t. Thus, we assign:

$$\beta_t = 1, \ t = 1, 2, \ldots, T$$

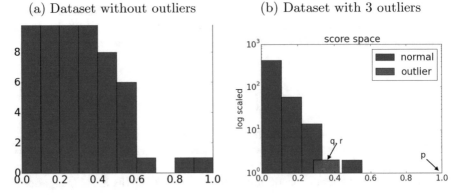

Fig. 8.3 Difference between the histograms of scores, **(a)** without anomalies; and **(b)** with three anomalies

b) Another approach is to elect a score threshold τ, and at each iteration t, calculate a_t =the number of objects with score greater than τ. Using this, we assign:

$$\beta_t = 1 - \frac{a_t}{|\mathscr{D}_t|}$$

where $|\mathscr{D}_t|$ is the size of the sample in the tth iteration.

c) At each iteration t, we may instead obtain the histogram of the score output, calculate the 'gap' b_t between the right-most bin and the second right-most bin, and set:

$$\beta_t = \frac{b_t + 1}{\sum\limits_{t=1}^{T} b_t + 1}.$$

8.4.1 Weighted Adaptive Sampling Algorithm

Algorithm "Anomaly detection with weighted adaptive sampling" begins by giving equal sampling weight to all points in the original dataset $\mathscr{D}$, such that for every point x_i, $w_{x_i}^0 = 1/|\mathscr{D}|$. At each iteration $t \in \{1, 2, \ldots, T\}$, we draw N observations following the sampling distribution $\mathbf{p}^t$, obtained as described below. Duplicates are removed and the scores are re-evaluated, and this new set of observations is denoted as $\mathscr{D}_t$. We adjust the sampling weights for all the points in $\mathscr{D}_t$ as mentioned above, and normalize their sampling weights with respect to the sum of all the sampling weights in $\mathscr{D}_t$; for unsampled data, the sampling weights remain unchanged. The result of our sampling makes the sampling weights for possible boundary points (hard-to-identify points) higher in the following iterations, so the 'effective' sample size will decrease over successive iterations.

8.4.2 Comparative Results

This section describes the datasets used for simulation, simulation results, and discussions of algorithm parameters.

Many researchers use the *Receiver Operating Characteristic (ROC)* curve, plotting true positive rate against false positive rate, to show how the performance of a classifier varies as its threshold parameters change. The *Area Under Curve (AUC)* measure, defined as the surface area under ROC curve, is used here to evaluate the relative performance of anomaly detection algorithms. A larger value of AUC indicates that the algorithm is more capable of capturing anomalies while introducing fewer false positives.

Algorithm Anomaly detection with weighted adaptive sampling

Input: Dataset $\mathscr{D}$, and detection algorithm A.
Initialize the weight vector: $w^0_{x_i} = \frac{1}{|\mathscr{D}|}$ for each point x_i;
for iteration t = 1,2,...,T:
1. Obtain the probability distribution for sampling:

$$\mathbf{p^t} = \frac{\mathbf{w^t}}{\Sigma w^t_{x_i}}$$

2. Draw N observations from $\mathscr{D}$ using $\mathbf{p^t}$, remove the duplicates, denote this set of observations as $\mathscr{D}_t$
3. Run A on $\mathscr{D}_t$, get a score vector $\mathbf{Score^t}$, normalize the scores to [0,1]
4. $h_t(x_i)$= normalized score of x_i
5. Set the new weights vector to be:

$$w^{t+1}_{x_i} = (1 - \alpha) * w^t_{x_i} + (\alpha) * WB(x_i);$$

Output: Make the final decision

$$H(x_i) = \sum_{t=1}^{T} \beta_t h_t(x_i)$$

8.4.3   Dataset Description

Five well-known datasets were used in our simulations; described in the Appendix:

1. Packed Executables Classification dataset (PEC)
2. KDD 99 dataset (KDD)
3. Wisconsin Dataset (WIS)
4. Basketball dataset (NBA)
5. Smartphone-Based Recognition of Human Activities and Postural Transitions Data Set (ACT)

8.4.4   Performance Comparisons

The ensemble approach was compared with base algorithm LOF using different values of k (the number of local neighbors). Figures 8.4 and 8.5 plot AUC values against different values of k. The solid line shows the AUC for base algorithm LOF, and the boxplots show the results over 20 trials of the ensemble approach with adaptive weighted sampling, using $\beta_t = 1 - \frac{a_t}{|D_t|}$. The results show that for all k values, the ensemble approach outperforms the base algorithm for all the 5 datasets; we observe that the variance in performance quality decreases as k increases.

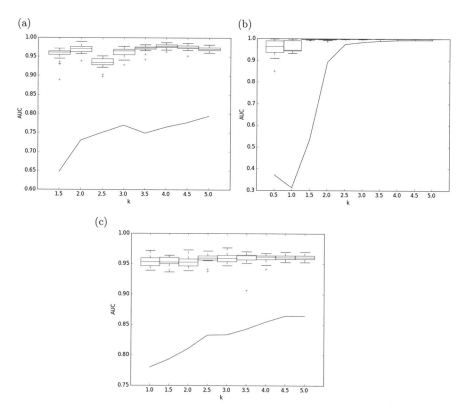

Fig. 8.4 AUC performance comparison of weighted adaptive sampling (boxplots) with base algorithm LOF (*solid curve*), for different values of k. (**a**) Wisconsin dataset. (**b**) Activity dataset. (**c**) NBA dataset

In the Active-Outlier approach [1], the anomaly detection problem is transformed to a classification problem by inserting additional observations generated from a specific distribution (Uniform, Gaussian). The inserted observations are labeled to belong to the anomalous class, and the observations in the original dataset are assumed to be non-anomalous. AdaBoost may be used with decision tree classifier (CART) as the base classifier. In the ensemble approach, no additional observations are inserted; LOF is used as the base anomaly detection algorithm with $\alpha = 0.2$ and results are reported for three different values of k. For fair comparison, the same number of iterations are performed with both methods.

Table 8.1 presents the AUC results, averaged over 20 trials; the best values are shown in boldface. The relative performance of the ensemble approach is better in all but one cases and improves as k increases. On the mobile activity dataset, the Active-Outlier approach fails to detect any real anomalies when a uniform distribution is used for synthetic anomaly generation, but performs significantly better when a Gaussian distribution is used.

Fig. 8.5 Comparison of
AUC performance: weighted
adaptive sampling (boxplots)
versus the base algorithm,
LOF (*solid lines*); the x-axis
shows the percentage of the
data set size considered to be
local neighbors: (**a**) KDD
dataset; (**b**) PEC dataset

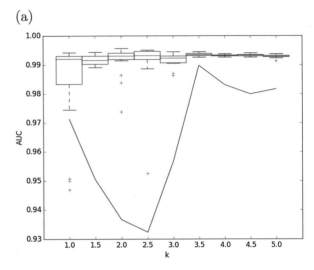

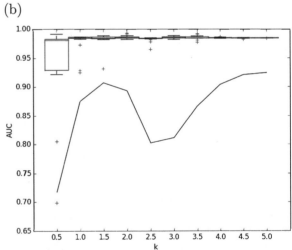

Table 8.1 Performance
comparison: averaged AUC
over 20 repeats

DataSet	Active-Outlier		Ensemble		
	Unif	Gauss	k=1%	k=3%	k=5%
WIS	0.865	0.921	0.963	0.965	**0.978**
NBA	0.903	0.810	0.920	0.959	**0.961**
ACT	0.5	0.980	0.961	0.999	**1.000**
PEC	0.907	0.686	0.975	**0.986**	0.985
KDD	0.873	0.685	0.985	**0.993**	0.993

8.4.5 Effect of Model Parameters

We first discuss how the number of iterations affects the performance of the ensemble approach. Figure 8.6 plots the AUC values against the number of iterations for a synthetic dataset and a real dataset. On the synthetic dataset, performance stabilized after 10 iterations and on the real dataset, performance stabilized after 20 iterations.

The simulations for comparing different possible combination approaches were conducted as follows:

1. For each dataset, for each k value, the algorithm is applied with each of the three different combination approaches.
2. The above process is repeated 20 times, and the mean AUC is reported.
3. For each dataset, for each k value, the AUC performance of each combination approach is ranked; the best one will have a rank of 1, the worst one has a rank of 3.
4. For each dataset, the sum of the above ranks is calculated for each combination approach; the best one has the lowest sum.

Table 8.2 summarizes the accumulated sums of AUC rankings over all different k values for the three approaches, denoted as Sum_a, Sum_b, and Sum_c respectively; the best values are shown in boldface. The one with best detection performance will

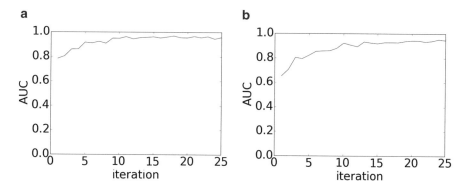

Fig. 8.6 AUC performance vs. number of iterations. (**a**) Synthetic dataset. (**b**) Real dataset

Table 8.2 Performance over Sum of AUC rankings with different combination approaches

DataSet	Sum_a	Sum_b	Sum_c
WIS	42	**33**	73
NBA	55	52	**43**
ACT	41	**27**	73
PEC	59	45	**33**
KDD	**33**	41	57
Sum over all datasets	230	**198**	279

have the lowest *Sum* over the three different combination approaches. Approach *a* ranks once as the best, twice as the second, and twice as the third; combination approach *c* ranks twice as the best, three times as the third; while the combination approach *b* ranks twice as the best, three times as the second and none as the third. So the combination approach *b*, where $\beta_t = 1 - \frac{a_t}{|D_t|}$, outperforms the other two combination approaches over the 5 datasets that were considered.

8.5   Conclusion

This chapter has presented several anomaly detection algorithms based on ensemble approaches. First, the independent ensemble approach provided algorithms for combining the results after separately applying each individual algorithm. Then, the sequential ensemble approach suggests how the results of applying one algorithm can be refined by the application of another algorithm. We then considered adaptive sampling methods, building on the well-known Adaboost algorithm, including a discussion of active learning in this context. Finally, weighted adaptive sampling was presented and evaluated using five datasets, leading to the conclusion that such an approach outperforms other algorithms discussed earlier.

Chapter 9
Algorithms for Time Series Data

Many practical problems involve data that arrive over time, and are hence in a strict temporal sequence. As discussed in Chap. 5, treating the data as a set, while ignoring the time-stamp, loses information essential to the problem. Treating the time-stamp as just another dimension (on par with other relevant dimensions such as dollar amounts) can only confuse the matter: the occurrence of other attribute values at a specific time instant can mean something quite different from the same attribute values occurring at another time, depending on the immediately preceding values. Such dependencies necessitate considering time as a special aspect of the data for explicit modeling, and treating the data as a sequence rather than a set. Hence anomaly detection for time-sequenced data requires algorithms that are substantially different from those discussed in the previous chapters.

Applications of time series anomaly detection have emerged in many fields, including health care, intrusion detection, finance, aviation, and ecology. For example, in a computer network, an anomalous traffic pattern may be observed when a hacked computer sends sensitive data to an unauthorized destination. In health management, an abnormal medical condition in a patient's heart can be detected by identifying anomalies in the time-series corresponding to Electrocardiogram (ECG) recordings of the patient; a normal ECG recording produces a periodic time series, whereas the ECG of a patient exhibiting arrhythmia may contain segments that do not conform to the expected periodicity or amplitude. In the domain of aviation, flight data is collected in the form of sequences of observations from multiple sensors, resulting in a multi-variate time series; any significant deviation from the typical system behavior is considered anomalous. Anomaly detection has been used in astronomical data to detect light curves' outliers within astronomical data [96]. Outlier detection for temporal data has been extensively surveyed by [23, 52].

This chapter is organized as follows. In Sect. 9.1, we describe the main anomaly detection problems generally addressed in time series context. In Sect. 9.2, we describe algorithms for identifying anomalous sequences from a collection of sequences, including a discussion of distance measures and popular transformations

K.G. Mehrotra et al., *Anomaly Detection Principles and Algorithms*, Terrorism, Security, and Computation, https://doi.org/10.1007/978-3-319-67526-8_9

that have been used in developing the anomaly detection algorithms. In Sect. 9.3, we consider algorithms for identifying anomalous subsequences. All of these algorithms are based on a single distance measure; in Sect. 9.4, we consider detection algorithms based on combining multiple distance measures and evaluation procedures. Online time series anomaly detection is discussed in Sect. 9.5. Some of these algorithms are empirically evaluated and compared; results are presented in Sect. 9.6. A summary of these algorithms and a discussion of the results are contained in Sect. 9.7.

9.1 Problem Definition

Different kinds of anomaly detection problems are illustrated by the example of financial time series shown in Fig. 9.1, displaying the variation of stock prices of some companies for the period from 2010 to 2012. First, we observe that one time series (Walmart Inc., shown with red dots) exhibits behavior substantially different from others, which represent the stock prices of oil and gas companies. Within each series, there are considerable fluctuations, among which some sudden drastic decreases stand out as anomalies, perhaps indicating problems of a specific company. At some time points, indicating market crashes, all the time series exhibit substantial decreases: the decrease of one company's stock price then is anomalous with respect to its own past behavior, but not with respect to the overall market behavior, which could be represented by a time series that aggregates multiple companies (e.g., Dow Jones Industrial Average, or the S&P500 and similar indices).

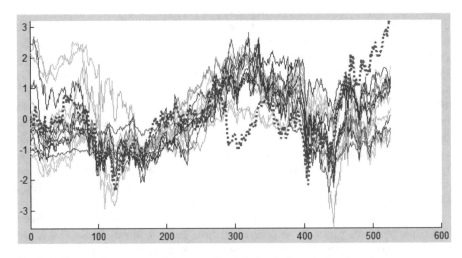

Fig. 9.1 Time series representing normalized daily closing stock prices for some companies—*Red dots* represent the stock prices of Walmart, Inc., whose behavior varies substantially from that of others representing various oil and gas companies

We hence discuss both kinds of anomaly detection algorithms: distinguishing an anomalous sequence from others, as well as identifying anomalous subsequences within a given sequence; furthermore, online anomaly detection algorithms are necessary in order to react rapidly to market fluctuations. These problems are discussed below in greater detail.

- **Between various time series:** We first consider the problem of identifying which time series is anomalous when compared to others in a collection of multiple time series.

  1. First, we consider the entire time period for which data are available, e.g., determining that the stock prices of Walmart represent an outlier with respect to the oil and gas stock price time series, shown in Fig. 9.1. Formally, suppose we are given a time series dataset $\mathscr{D} = \{\mathscr{X}_i | i = 1, 2, \ldots, m\}$, where $\mathscr{X}_i = \{x_i(t) | 1 \leq t \leq n\}$ represents the ith time series and $x_i(t)$ represents the value of the ith time series at time t, n is the length of the time series, and m is the number of time series in the dataset. The goal is to calculate $O(\mathscr{X}_i)$, the outlierness or anomaly score[1] of a series $\mathscr{X}_i$, and Θ_0 a threshold such that if $\mathscr{X}_i$ is an anomalous series, then $O(\mathscr{X}_i) \geq \Theta_0$; otherwise $O(\mathscr{X}_i) < \Theta_0$.
  2. Although a time series may resemble other time series in the dataset when the entire period of time is considered, it may diverge substantially from the others for a small sub-period. The change may be permanent or short-lived; the behavior of such a company stock price may revert later to its previous behavior, once again moving in tandem with other stocks in the same industry group. For example, among the stock prices of various oil and gas companies, which mostly move in tandem, one particular stock may exhibit anomalous behavior, e.g., rising or dipping suddenly due to rumors of a possible merger. On the other hand, a company's stock price may be almost unchanged over time, but this may be anomalous when compared to the stock prices of other companies, all of which exhibit sudden spurts in response to external news (e.g., impending war that increases demand for some commodities). Adding to the notation mentioned earlier, we must calculate $O(\mathscr{X}_i, p, w, \mathscr{D})$ which represents the outlierness of the time series $\mathscr{X}_i$ during the time interval from time instant p to $p + w$, compared to other time series $\in \mathscr{D}$ for the same time period. The goal is to find the series $\mathscr{X}_i$ as well as the relevant time period parameters (p, w) such that $O(\mathscr{X}_i, p, w, \mathscr{D})$ exceeds a threshold. This requires very high computational effort, unless the possible values of p and w are constrained substantially.

- **Within a time series:** Another problem of interest is the detection of an anomaly within a single series, which differs substantially from the rest of the same time series; identifying the precise moment when the deviation begins could be important in some cases.

---

[1] We use "O" instead of "a" in the case of time series; these are otherwise identical except that there is no presumption that $O \leq 1$.

1. The occurrence of an *event* or *point anomaly* is characterized by a substantial variation in the value of a data point from preceding data points. For example, in the credit card fraud detection context, if a purchase is associated with a much higher cost than prior purchases by the same customer, an anomaly would be signaled.

2. *Discords* are anomalous subsequences that occur within a given time series, defined as "maximally different from the rest of the sequence" [72, 73]. For example, in detecting cardiac arrhythmias from an electrocardiograph (ECG), no individual reading may be out of range, but the sequence of a collection of successive values may not be consistent with the regular and periodic form of the data. We expect a certain (previously observed) regular waveform to be repeated, and variation from this expectation constitutes an anomaly. Another example consists of overseas money transfers; past history might indicate that these occur about once a month for a customer (e.g., to financially support the customer's family living abroad), representing a simple periodic sequence over time; the occurrence of multiple such transactions within a month then constitutes an anomaly. Detecting such anomalies is difficult because the exact length of the anomalous subsequence may not be known.

   Formally, suppose $\mathcal{X} = \{x(t)|1 \leq t \leq n\}$ is a time series where $x(t)$ represents the value of the series $\mathcal{X}$ at time t and n is the length of $\mathcal{X}$. A subsequence of $\mathcal{X}$ is defined as $\mathcal{X}_{p,w} = \{x(p), \ldots, x(p + w - 1)|1 \leq p \leq n - w + 1; 1 \leq w \leq n\}$. The goal is to find abnormal subsequences ($\mathcal{X}_{p,w}$) if any exist. More specifically, we must calculate the outlierness, $O(\mathcal{X}_{p,w})$, of any $\mathcal{X}_{p,w}$. If $O(\mathcal{X}_{p,w}) \geq \Theta_0$, a user-defined threshold, then $\mathcal{X}_{p,w}$ is declared to be abnormal.

3. Sometimes the individual values may be within an acceptable range, but the rate of change over time may be anomalous, and we refer to this as a *rate anomaly*. For instance, in the credit card fraud detection problem, the balance (total amount owed by the customer) may suddenly increase within a day, due to a large number of small transactions made during a short amount of time; this should signal an anomaly even if each individual transaction is unremarkable, and the total balance remains within the range of prior monthly expenditures. This is an example of a *contextual anomaly,* wherein a data point is anomalous with respect to the immediately preceding values, though not with respect to the entire range of possible values from past history.

Example 9.1 The rapid fall of an oil stock price may signal an anomaly when viewed against its own past history. But this may not be an anomaly when evaluated against simultaneous rapid decreases in other oil stock prices. On the other hand, the stock price of an oil company may at some time begin to exhibit significant deviations from those of other oil companies, suggesting an underlying problem that should trigger actions by stock traders. In order to carry out such actions, it is important to detect such deviations as they occur in time, else the trader may be too late and may suffer losses due to delays in analysis and decision-making.

- **Online anomaly detection:**   In some time series problems, the underlying process generating the data changes with time. If we have applied a learning algorithm to determine the parameters of a model using old data, then the old parameters may no longer be applicable to new data. Hence we must reapply the learning algorithm to obtain a new set of parameters for the model. Unfortunately, this requires exorbitantly high computational costs, due to repeated applications of a learning algorithm when new data arrive. This makes it very difficult to detect anomalies as soon as they arise, in order to respond rapidly when anomalies are detected. This is referred to as the *online anomaly detection* problem. The goal may be to identify anomalies within a given series, but may also refer to anomalies in comparison to the most recent behavior of other time series.

  Formally, consider a time series dataset $\mathscr{D} = \{\mathscr{X}_i | i = 1, 2, \ldots, m\}$, where $\mathscr{X}_i = \{x_i(t) | t = 1, 2, \ldots\}$ represents the ith time series and $x_i(t)$ represents the value of the ith time series at time t. In the online context, the goal is to calculate $O(x_i(t))$, the outlierness of a series x_i at time t, without requiring excessive computation but learning the characteristics of the most recent data, i.e.,

$$\{(\mathscr{X}_j(t - w), \ldots, \mathscr{X}_j(t - 1)), \text{ for } j = 1, \ldots m\}.$$

The appropriate window length w is itself unknown, and may depend on the problem characteristics, model size, prior hypotheses regarding the possible rate of change of the underlying process, and inferences made from data.

Problems in which available data are not evenly spaced over time constitute another challenge. For example, during a single day, trades in a stock may occur at uneven intervals. In other cases, data at all time points may exist but may be unavailable or missing, e.g., when sensors take readings at varying intervals in time. In observations of physical data (e.g., temperature at the bottom of a lake), for instance, available data may constitute a discrete irregularly spaced sample of observations from an underlying infinite data set that varies continuously over time. Interpolation or imputation methods, [9], may sometimes be applied to infer missing data, prior to the application of anomaly detection algorithms of the kind discussed in this chapter.

Another problem arises when all time series are not necessarily of the same length; for presentational convenience, we ignore this concern.

9.2   Identification of an Anomalous Time Series

Just as earlier chapters discussed the identification of anomalous data points (outliers) from other data points (inliers), the classic problem for time series is to identify the time series that is an outlier with respect to other time series. Abstractly, algorithms similar to those discussed in previous chapters can again be applied if we can identify a space in which time series can be compared easily with each other, and an appropriate distance measure for time series. However, these are non-trivial issues, and several approaches have been explored to address the relevant task. We first describe various categories of algorithms used for this task, and then discuss the distance measures and transformations that are specific to time series problems.

9.2.1 Algorithm Categories

Anomaly detection algorithms for data sequences fall in two major categories, *viz.*, Procedural and Transformation approaches.

In procedural or model-based algorithms a parametric model, such as regression or Hidden Markov Models (HMMs), is built using the training data to predict the behavior of the time series, and an anomaly score is assigned to each (test) time series. The predicted values are calculated and compared with the observed values, as sketched in Algorithm "Procedural approach"; note that we may choose the testing period, which need not necessarily begin after the training period, i.e., it is possible that $\tau' < \tau + 1$.

Algorithm Procedural approach

1: GIVEN: Time series dataset $\mathcal{D}$, training time period τ, testing start point τ', parametric model M
2: Train the model M on the first τ values of time series in $\mathcal{D}$ to determine appropriate parameter values, and let the result be M_τ;
3: **for** each time series $\mathcal{X}_i \in \mathcal{D}$ **do**
4:     Apply M_τ to predict $x_i(\tau'), \ldots, x_i(n)$; thus obtain $\hat{x}_i(\tau'), \ldots, \hat{x}_i(n)$;
5:     Report $\mathcal{X}_i$ as an anomaly if $(\hat{x}_i(\tau'), \ldots, \hat{x}_i(n))$ is substantially different from $(x_i(\tau'), \ldots, x_i(n))$;
6: **end for**

The models used for this approach include Regression [43, 101], Auto-Regression [45], ARMA [95], ARIMA [88], and Support Vector Regression [83, 106]. These methods are mainly designed for individual outlier detection, and the results are impacted significantly by model parameters or distributions of data sets.

In the transformation approach, the data is transformed prior to anomaly detection, using approaches such as the following:

- *Aggregation approach* which focuses on dimensionality reduction by aggregating consecutive values; this approach is sketched in Algorithm "Aggregation approach".
- *Discretization approach* which converts numerical attributes into a small number of discrete values, in order to reduce computational effort; this approach is sketched in Algorithm "Discretization approach".
- *Signal processing transformations*, e.g., in Fourier or Wavelet transformation the data is transformed to a different space; a subset of transformed parameters is chosen, thereby reducing the dimensionality. This approach is sketched in Algorithm "Signal processing transformation approach".

Algorithm Aggregation approach

1: GIVEN: Time series dataset $\mathscr{D}$, window size w (much smaller than the time series length);
2: **for** each time series $\mathscr{X}_i \in \mathscr{D}$ **do**
3: **for** $j \in [1, \ldots, \lfloor \frac{n}{w} \rfloor]$ **do**
4:   Aggregate w consecutive values
    $A_i[j] = (x_i(j * w + 1), x(j * w + 2), \ldots, x(j * (w + 1)))$;
5: **end for**
6: **end for**
7: Apply an anomaly detection algorithm to the aggregated time series $\{A_i, i = 1, 2, \ldots, m\}$ where $A_i = (A_i[1], A_i[2], \ldots, A_i[\lfloor \frac{n}{w} \rfloor])$

Algorithm Discretization approach

1: GIVEN: Time series dataset $\mathscr{D}$, discrete value count v (often in the 3–5 range);
2: **for** each time series $\mathscr{X}_i \in \mathscr{D}$ **do**
3: **for** $j \in [1, n]$ **do**
4:   Discretize $x_i(j)$ into one of the v values, so that $v_i(j) \in \{1, \ldots, v\}$;
5: **end for**
6: **end for**
7: Apply an anomaly detection algorithm to the discretized time series $\{V_1, V_2, \ldots\}$ where $V_i = (v_i(1), v_i(2), \ldots, v_i(\lfloor \frac{n}{w} \rfloor))$

Algorithm Signal processing transformation approach

1: GIVEN: Time series dataset $\mathscr{D}$, Transformation F, Desired dimensionality d (often in the 3–6 range);
2: **for** each time series $\mathscr{X}_i \in \mathscr{D}$ **do**
3: Compute $F(\mathscr{X}_i)$;
4: Apply dimensionality reduction to obtain $F^*(\mathscr{X}_i)$ from $F(\mathscr{X}_i)$, e.g., by selecting the d most significant coefficients of terms in $F(\mathscr{X}_i)$;
5: **end for**
6: Apply an anomaly detection algorithm to the transformed data $\{F^*(\mathscr{X}_1), F^*(\mathscr{X}_2), \ldots\}$

We now address the formulation of distance metrics and transformations that are central to these approaches.

9.2.2 Distances and Transformations

Viewed as a vector, a time series is generally of very large dimensionality. Consequently, the first step is to obtain a compact representation to capture important information contained in the series. Lin et al. [79] define a generic time series data mining approach as follows:

1. Create an approximation of the data, which will fit in main memory, yet retains the essential features of interest.
2. Approximately solve the task at hand in main memory.

3. Make a small number of accesses to the original data on disk to confirm the
   solution obtained in Step 2, or to modify the solution so it agrees with the solution
   we would have obtained on the original data.

Some approaches to reduce the dimensionality of time series are model-based,
whereas others are data-adaptive [20, 33].

The *lock-step* measures are based on one-to-one mapping between two time
series; examples include: Cross Euclidean distance (EUC) [40], the Cross Correlation Coefficient-based measure [13], SameTrend (STREND) [60], and Standard
Deviation of Differences (DIFFSTD), discussed below.[2]

- The *Cross-Euclidean distance* between two time series $\mathcal{X}$ and $\mathcal{Y}$ is defined as:

$$d_E = \left(\sum_{t=1}^{n} (x(t) - y(t))^2 \right)^{\frac{1}{2}}.$$

- Better scale invariance is provided by the *Cross Correlation Coefficient*-based
  measure, defined as

$$d_C = (2(1 - \text{correlation}(\mathcal{X}, \mathcal{Y})))^{\frac{1}{2}},.$$

  where correlation is the Pearson correlation coefficient between the two time
  series.
- For the *SameTrend (STREND)* measure, the difference $\Delta x(t) = x(t+1) - x(t), t \in$
  $[1..n-1]$ is calculated for each series, and STREND is defined as

$$S(t) = \begin{cases} 1 & \text{if } \Delta x(t) \cdot \Delta y(t) > 0 \\ -1 & \text{if } \Delta x(t) \cdot \Delta y(t) < 0 \\ 0 & \text{otherwise.} \end{cases}$$

Thus, $S(t)$ indicates whether or not $x(t)$ and $y(t)$ change in the same direction
at time t. The aggregate measure, over the entire length, n, of the time series is
evaluated as

$$d_S(\mathcal{X}, \mathcal{Y}) = 1 - \frac{1}{n-1} \sum_{t \in [1..n-1]} S(t).$$

Example 9.2 Figure 9.2 illustrates the STREND computation for two time
series. For time series $\mathcal{X}$, the successive values are $(9, 3, 4, 5, 3, 5, 6, 2, 1)$,
and the differences (subtracting each element from the preceding one)

---

[2]For readability, we often abbreviate the aggregate distance between two time series $\mathcal{X}_i$ and $\mathcal{X}_j$ as
$d(i, j)$ or $dist(i, j)$ instead of $d(\mathcal{X}_i, \mathcal{X}_j)$.

Sign(x(t+1)–x(t))

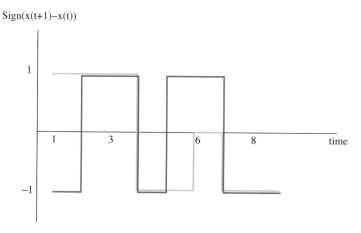

Fig. 9.2 Plots of trend ($sgn(\Delta x(t) = x_{t+1} - x_t)$) for two time series. These two series move in the same direction five out of eight times

are $(-6, 1, 1, -2, 2, 1, -4, -1)$. Similarly, for time series $\mathcal{Y}$, the values are $(1, 2, 3, 4, 3, 2, 2, 1, 0)$ so that the successive differences are $(1, 1, 1, -1, -1, 0, -1, -1)$. The trend for the first interval is negative for series $\mathcal{X}$ and positive for series $\mathcal{Y}$, hence the first element of the STREND vector is $S(1) = -1$; but the trends are both positive for the second interval, hence the second element of the STREND vector is $S(2) = 1$. All the elements of the STREND vector are given by $S = (-1, 1, 1, 1, -1, 0, 1, 1)$, in which the sixth element is 0 since the corresponding trend is $S(6) = 0$ (neither positive nor negative) for $\mathcal{Y}$. Finally, the STREND distance (between $\mathcal{X}$ and $\mathcal{Y}$) computed is $1 - \sum_t (S(t))/(n-1) = 1 - (3/8) = 0.625$.

- DIFFSTD is the standard deviation of differences between two time series, i.e., if $\delta_{\mathcal{X}, \mathcal{Y}}(t) \equiv \delta(t) = \|x(t) - y(t)\|$, and $\mu = \sum_t \delta(t)/n$, then the new distance is defined as

$$dist(\mathcal{X}, \mathcal{Y}) = \sqrt{\sum_t (\delta(t) - \mu)^2/n}.$$

This measure, illustrated in Fig. 9.3, is widely used in the financial field.

Example 9.3 For the same two time series (used to illustrate STREND above), the DIFFSTD measures begins by computing the absolute difference $\delta(t) = \mathcal{X}(t) - \mathcal{Y}(t)$ between corresponding elements of the two time series, giving values $(8, 1, 1, 1, 0, 3, 4, 1, 1)$ from which we compute the average value $\mu = \sum_t \delta(t)/n = 2.22$ and finally the distance

$$\sqrt{\frac{\sum_t (\delta(t) - \mu)^2}{n}} = 2.35$$

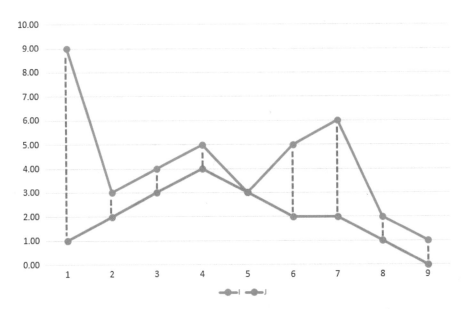

Fig. 9.3 Two time series compared using DIFFSTD—*Vertical lines* indicate differences at individual time points

- The elastic measures are based on one to many mapping between two time series and therefore can handle the time shifting/lag issues. An elastic measure matches one point of the time series versus many or none points of another time series, so that it always tries to find the best possible matches between two time series under certain conditions.
- In recent years, some time series researchers have applied the measures used to compare strings (finite sequences of characters), such as *Edit distance*, defined as the smallest number of additions, deletions, and insertions needed to make two strings identical. *Edit Distance on Real sequence* (EDR) [24] is a modification required since the elements of a time series may not be integers, and ignores differences between elements that are below a minimal threshold parameter ϵ, as follows:

$$\delta_\epsilon(x_i, y_i) = \begin{cases} 0 & \text{if } |x_i - y_i| < \epsilon \\ 1 & \text{otherwise} \end{cases}$$

For example, when $\epsilon = 0.15$, and insertions, deletions, and gaps are equally penalized, the edit distance between the sequences $x = (0.1, 0.5, 0.5, 0.8, 0.9)$ and $y = (0.3, 0.6, 0.8)$ is given by the alignment

$$((0.1, \_), \ (\_, 0.3), \ (0.5, 0.6), \ (0.5, \_), \ (0.8, 0.8), \ (0.9, \_))$$

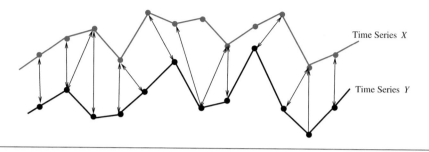

Fig. 9.4 Time alignment of two time series; aligned points are indicated by the *arrows*

which gives

$$\delta_{0.15}(x, y) = 1 + 1 + 0 + 1 + 0 + 1 = 4.$$

Variations of this approach include assigning greater penalties to gaps, especially those that occur in the middle of the sequence. When penalties are very large, the problem reduces to finding the longest common substring, and when gaps are not penalized, the problem reduces to finding the longest common subsequence (LCSS) [113]. A similar approach is used in *Dynamic time warping (DTW)* [11, 99], used in many practical applications such as speech recognition and signature recognition. DTW attempts to accommodate such discrepancies in $\mathscr{X}$ and $\mathscr{Y}$ series; *viz.* Fig. 9.4. A *warping path* for $\mathscr{X}$ and $\mathscr{Y}$ is defined as a pair of monotonically non-decreasing sequences $L = (1, \ldots, L_k, \ldots, \|\mathscr{X}\|)$ and $R = (1, \ldots, R_k, \ldots, \|\mathscr{Y}\|)$, representing indices into the two time series, respectively, constrained so that

$$(L_k - L_{k-1}, R_k - R_{k-1}) \in \{(0, 1), (1, 0), (1, 1)\}.$$

The cost associated with a warping path (L, R) is defined as $\sum_k d(\mathscr{X}(L_k) - \mathscr{Y}(R_k))$, where d measures the difference in component values. In DTW the goal is to find the warping path that minimizes this cost. This can be accomplished by using the classical dynamic programming algorithm developed to align two sequences [90].

- In the transformed space approach, the time series is first transformed into another space; measures in the transformed space include TQuEST [2] distance, and Spatial Assembling Distance SpADe [26]. Other examples include Symbolic Aggregate approXimation (SAX) proposed by Keogh and Lin [79] with and without sliding window (SAXSL and SAXNW respectively); SAX with bag-of-pattern (SAXBAG) [80], Discrete Wavelet Transform [19], and Discrete Fourier Transform [40]. We describe a few of these distances in the following.

- Piecewise Aggregate Approximation (PAA) is a simple dimensionality reduction method for time series mining. PAA approximates a time-series $\mathcal{X}$ of length n into a vector of length M where $M < n$. The transformed vector $\tilde{\mathcal{X}} = (\tilde{x}_1, \tilde{x}_2, \ldots, \tilde{x}_M)$ where each $\tilde{x}_i$ is calculated as follows:

$$\tilde{x}_i = \frac{M}{n} \sum_{\lfloor \frac{n}{M}(i-1)\rfloor+1}^{\frac{n}{M}(i)} x_j.$$

Thus, an n-dimensional time series is reduced to an M-dimensional vector which consists of the mean values for each $\frac{n}{M}$-dimensional subsequence of the series. The distance between two time series $\mathcal{X}$ and $\mathcal{Y}$ is approximated using the distance, d_P, between the PAA of each series where:

$$d_P(\tilde{\mathcal{X}}, \tilde{\mathcal{Y}}) = \sqrt{\frac{n}{M}} (\sum_{i=1}^{M} (\tilde{x}_i - \tilde{y}_i)^2)^{\frac{1}{2}}.$$

The PAA distance satisfies the bounding condition [119]:

$$d_P(\tilde{\mathcal{X}}, \tilde{\mathcal{Y}}) \leq \text{dist}(\mathcal{X}, \mathcal{Y}).$$

Note that a series $\mathcal{X}$ will be declared anomalous if its distance from other series is large. But, if in place of Euclidean distance the PAA distance is used, then the above inequality guarantees no false decision. A drawback of this measure is that PAA minimizes dimensionality by the mean values of equal sized subsequences, which misses some important information and sometimes causes inaccurate results in time series mining.

- SAX [79] converts a real-value univariate time series to a string by performing sliding window subsequence extraction with aggregation followed by discretization. First, it chooses a size of sliding window, then transforms all observations of the raw dataset inside the sliding window into Piecewise Aggregate Approximation (PAA [70]) representation; next it transforms PAA representation in the sliding window into a SAX word. After obtaining the SAX sequences of words, it is possible to perform optimization for the sequences containing consecutive repetitions of the same words, further reducing the length of the final output sequence, but we ignore this refinement in the following discussion.

  Suppose the size of the sliding window is w, and the alphabet Σ (used in SAX) is of size $|\Sigma|$. Then the following steps are executed:

  1. The time series $\mathcal{X}$ is transformed to a PAA sequence P as described earlier; $P = P(1), \ldots, P(M)$, where $M = \lceil \frac{n}{w} \rceil$.
  2. The sequence P is converted to a SAX word-sequence $S = S(1), \ldots, S(M)$, where $S(t) \in \Sigma$ and the associated value is determined by the magnitude of $P(t)$ in the equi-probability distribution. The equi-probability distribution identifies *breakpoints* $k_1, \ldots, k_{a-1}$ such that the area under the normal curve

in the interval $(-\infty, k_1)$ equals each area under the normal curve in the interval (k_i, k_{i+1}) and (k_{a-1}, ∞). When $a = 3$, for instance, $k_1 \approx -0.43$ and $k_2 \approx 0.43$, and each area under the normal curve in the intervals $(-\infty, -0.43)$, $(-0, 43, 0.43)$, and $(0.43, \infty)$ is approximately 1/3. Any value can be discretized into one of the three ranges into which the normal curve is divided by these two breakpoints (k_1, k_2), e.g., -1 would be discretized into the symbol representing the interval $(-\infty, -0.43)$, 0.5 into the symbol representing the interval $(-0.43, 0.43)$, and 2 into the symbol representing the interval $(0.43, \infty)$.

Thus, the time series $\mathscr{X}$ is represented as a sequence of SAX words $S(t), t = 1, \ldots, M$, as illustrated in Fig. 9.5.

3. The distance between two time series $\mathscr{X}, \mathscr{Y}$ is defined as

$$d_{SAX}(\mathscr{X}, \mathscr{Y}) = \left(\frac{n}{M} \sum_{t=1}^{M} symdist(S(t), S^*(t))^2 \right)^{\frac{1}{2}}.$$

where S^* represents the SAX word representation of time series $\mathscr{Y}$ and *symdist* is the symbolic distance that can be calculated using the *SAX distance lookup table* [79]. If $i < j$, then the *symdist* (table entry) between a symbol representing the ith interval (k_{i-1}, k_i) and a symbol representing the jth interval (k_{j-1}, k_j), is defined to be $k_{j-1} - k_i$, the difference between the lower bound of the higher interval and the upper bound of the lower interval. The $sym - dist$ table is symmetric, with 0 entries along the main diagonal and its two adjacent diagonals. For example, if three symbols (intervals) are used, the *symdist* between the extreme intervals is $0.43 - (-0.43) = 0.86$, and the other entries

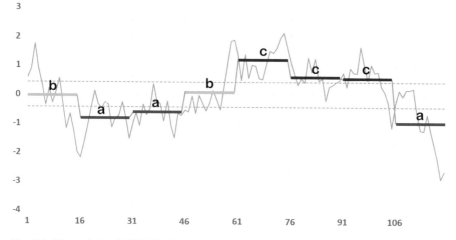

Fig. 9.5 Example for SAXBAG distance computation—SAX words and SAXBAG frequency counts

are 0. If four symbols are used, with three breakpoints $(-0.67, 0, 0.67)$, the symbol for the leftmost (smallest) interval is at a *symdist* of 0 from the second symbol, 0.67 from the third symbol, and 1.34 from the fourth (highest) symbol.

- **SAX with bag-of-pattern (SAXBAG):** This method uses the frequencies of different SAX words as the representation of the raw data [80]. Then the Euclidean distance can be calculated between two time series with bag-of-pattern representations. Given a time series, subsequences of size w are obtained using a sliding window. Each subsequence is reduced to w-dimensional discrete representation in two steps. First, in the aggregation step, (cf. Algorithm "Aggregation approach") the subsequence is further divided into $u = \frac{M}{w}$ *segments* (subsubsequences of equal size) and the average value of the data falling within a segment is evaluated. In the second step, each average is discretized to a symbol, from a predetermined set of symbols, using the equal probability intervals approach.

Example 9.4 Two time series are shown in Fig. 9.5. For concreteness, the values for one time series $\mathcal{X}$ are shown in Table B.1 in the Appendix (beginning with 0.00 and ending with -3.57). For this time series, the mean is -0.66, and the standard deviation is 1.10, and the z-scores obtained by subtracting the mean and dividing by the standard deviation, give the following results of normalization: 0.60, 0.88, 1.75, 0.91, 0.34, -0.38, 0.21, $\ldots$, -2.91, -2.64. We now use a window size of 15 and compute the averages of 15 successive datapoints per window (with no overlaps) to be -0.05, -0.81, -0.64, 0.10, 1.21, 0.59, 0.54, and -0.95, respectively. We use the alphabet size of 3, with three alphabet symbols $\in \{a, b, c\}$. The two cutpoints (from the equiprobable distribution) are at -0.43 and 0.43, dividing the range of values into three intervals; using these, we discretize the normalized sequence to the following: b, a, a, b, c, c, c, a. For example, the first symbol is b since $-0.43 \leq -0.05 \leq 0.43$, and the last symbol is a since $-0.95 < -0.43$. Using a word-size of 3, we obtain the following word pattern matrix; each row indicates the number of occurrences (in *baabccca*) of the listed 3-letter substring, but only for the non-zero cases. For example, *aaa* is omitted since there are no occurrences of *aaa* in *baabccca*, but *baa* 1 is listed since there is one occurrence of *baa* (at the beginning of the string *baabccca*).

$$aab : 1$$
$$abc : 1$$
$$baa : 1$$
$$bcc : 1$$
$$cca : 1$$
$$ccc : 1$$

In other words, the SAX representation of $\mathscr{X}$ will be $baa, aab, abc, bcc, ccc, cca$. Inserting 0 values for the absent substrings, and listing the values in alphabetical order, the full SAXBAG representation for time series $\mathscr{X}$ is:

$$0, 1, 0, 0, 0, 1, 0, 0, 0, 1, 0, 0, 0, 0, 0, 0, 0, 1, 0, 0, 0, 0, 0, 0, 1, 0, 1$$

where the first 0 corresponds to the (absent) alphabetically first substring aaa, followed by 1 for the next substring aab that does occur once in $baabccca$, etc. Similarly, the SAXBAG representation for another time series is as follows:

$$\mathbf{1}, 0, 0, 0, \mathbf{1}, \mathbf{0}, 0, 0, 0, 1, 0, \mathbf{1}, 0, 0, 0, 0, 0, 1, 0, 0, 0, 0, 0, 0, \mathbf{0}, 0, 1$$

where the values varying from the SAXBAG of $\mathscr{X}$ are shown in boldface.

Focusing on the values that are different between the two SAXBAGs, the SAXBAG distance between $\mathscr{X}$ and $\mathscr{Y}$ is finally computed to be:

$$\sqrt{((0-1)^2 + (1-0)^2 + (0-1)^2 + (1-0)^2 + (0-1)^2 + (1-0)^2)} = 2.45$$

- Inspired by Kolmogorov's complexity measure, Keogh et al. [74] defined the following dissimilarity measure:

$$d(x, y) = \frac{C(xy)}{C(x) + C(y)},$$

where $C(\ldots)$ denotes the compressed size of the argument string, and xy is the concatenation of x and y; the compression could be achieved using any suitable method, e.g., SAX.

9.3 Abnormal Subsequence Detection

Detecting anomalous subsequences involves evaluating whether a subsequence is substantially different from other subsequences (in the larger time series), and is anologous to evaluating whether an entire time series is substantially different from other time series, as indicated earlier. Hence the approaches, distance measures, and transformations mentioned in the previous section can once again be applied to the task of detecting anomalous subsequences.

One approach is to consider subsequences as points in space and use an existing anomaly detection approach to find anomalous subsequence(s). This approach is sketched in Algorithm "Point-based approach".

But the time complexity of this approach can be prohibitively high; a significant body of research to find anomalies in time series context addresses techniques to reduce this time complexity, e. g., using indexing techniques [71]. The main data structures that have been used towards efficient representation of subsequences

---

Algorithm Point-based approach

1: GIVEN: Time series X, window size w (much smaller than the time series length);
2: **for** each $i \in \{1, 2, \dots, (1 + \|X\| - w)\}$ **do**
3:     Let X_i be the subsequence of length w beginning with $X[i]$;
4: **end for**
5: Apply   an   anomaly   detection   algorithm   to   the w-dimensional   data   points $\{X_1, X_2, \dots, X_{(1+\|X\|-w)}\}$.

---

in 'feature' space are *R-tree, R*-tree and X-tree* representations. *R*-trees are data structures that are used for indexing multi-dimensional information such as geographical coordinates, rectangles or polygons.

Another approach defines anomalies within time series as occurrences of patterns that do not occur in normal circumstances, e.g., a flat section within a time series representing human heartbeats [67]. Features such as DWT coefficients are first extracted on sliding window sections of the time series, and these are discretized, using a single symbol to represent an interval of values. A data structure such as a suffix tree is then used to represent the frequencies of occurrence of various short strings of discretized features within the time series. Anomaly scores can then be computed by comparing suffix trees corresponding to different subsequences; see Algorithm "Wavelet-based Subsequence Anomaly Detection".

Algorithm Wavelet-based Subsequence Anomaly Detection

Require: Time series $\mathscr{X}$, Number of discretized values (v), Window length (w), and String length (l).
Ensure: outliers in $\mathscr{X}$.
Apply DWT to sliding window sections of the time series, where the ith window is $X(i) = (x(wi + 1), x_i(wi + 2), \dots, x_i(wi + w - 1))$.
Discretize DWT coefficients; each symbol $\in \{a_1, \dots, a_w\}$ represents values within an interval;
Compute the frequency $v_s^{(i)}$ in each window $X(i)$ of each substring $s = a_j, a_k, \dots$ of length l;
Represent these substring frequencies using a suffix tree;
Use the suffix trees to evaluate the distance between any two windows, measured in terms of the differences in substring frequencies, $d_v(X(i), X(j)) = \sum_s |v_s^{(i)} - v_s^{(j)}|$
Use the distance to compute the anomaly score of each window, e.g., as the distance to the kth nearest neighbors.

A similar algorithm compares the frequencies of SAX words of current data and past data [115].

With such an approach, if the length w of the discord (anomalous subsequence) is known, disjoint subsequences of length w can be compared to the entire sequence to find the anomalous subsequences. But no clear guideline exists for choosing the right values for window length; this problem exists for other algorithm parameters as well. One possibility is a 'Divide and Conquer' approach: repeatedly halve the sequence size until the right size is discovered, using a measure such as SAX compression dissimilarity (C), discussed earlier. Attention is focused on the part of a sequence which contains greater dissimilarity with the entire sequence, and the length of this subsequence is successively halved [74].

Table 9.1 summarizes advantages and disadvantages of various distance measures for anomaly detection in time series.

9.4   Outlier Detection Based on Multiple Measures

As indicated by Table 9.1, no single measure is capable of capturing different types of perturbations that may make a series anomalous; a similar observation was made to motivate ensemble methods in Chap. 8. Hence multiple measures should be considered together to improve anomaly detection for time series problems [60].

9.4.1   Measure Selection

For the best utilization of limited computational resources, we need to select a subset of the measures described in the previous section. Towards this goal, selection of measures that are orthogonal to each other is required, to minimize redundancy. The orthogonality can be approximated by selecting measures that are least correlated with each other; the first two measures are selected such that they are least correlated with each other and subsequent measures are selecting using partial correlation coefficients. Correlation and partial correlation coefficients between the measures can be computed over multiple datasets. Three useful measures that can be combined effectively are DIFFSTD, SAXBAG, and STREND [60].

These measures capture different aspects of the time series, and a combination of these gives a comprehensive measure of how isolated is a time series from others in the comparison set. SAXBAG captures the behavior of a time series using a histogram of possible patterns, STREND identifies the degree of synchronization of a series compared with another series, and DIFFSTD measures the amount of deviation. From another perspective, SAXBAG addresses the signature of a single time series, whereas DIFFSTD and STREND are inter-time-series-measures; STREND focuses on the synchronization of two time series whereas DIFFSTD focuses on the magnitude. Combining these three metrics produces a comprehensive and balanced measure that is more sensitive than individual measures.

Normalization After calculating these measures, they need to be normalized before combining, since the ranges of selected measures are very different; for example, the range of STREND is from 0 to 1, and the range of SAXBAG is from 0 to $l\sqrt{2}$ where l = the number of words obtained from the entire series. Empirical results have shown that the normalization method where each observation is divided by the trimmed mean (excluding 5% on either end) was found to perform better than

Table 9.1 Pros and Cons of using various distance measures for anomaly detection in time series

Measures	Pros	Cons
EUC[40] CCF [13] DIFFSTD [60]	Easy to implement; computationally efficient	Lock-step measure; normal series with time lagging cause problems
DTW [68]	Elastic measure; successfully addresses problems with time lagging between sequences	Small deviations may not be detected
DFT [40]	Good in detecting anomalies in frequency domain; normal series with time lagging can be solved	Small deviations may not be detected; cannot detect anomalies with time lagging
DWT [19]	Good in detecting anomalies in frequency domain;	Small deviations may not be detected; sensitive to time lagging
(SAXSL) SAX with sliding window [79]	Tolerates noise, as long as its standard deviation is small	May not detect abnormal subsequence of shorter length than feature window size; normal series with time lagging can result in false positives
(SAXNW) SAX without sliding window [79]	Tolerates noise, as long as its standard deviation is small;	May not detect abnormal subsequence of shorter length than feature window size; small deviations may not be detected; normal series with time lagging can result in false positives
(SAXBAG) SAX with bag-of-pattern[80]	Tolerates noise, as long as its standard deviation is small; normal series with time lagging can be solved	Cannot detect anomalies with time lagging; cannot detect anomalous series with similar frequencies but different shapes

alternatives that include the extreme values.[3] The normalized distances between the ith and the jth series, based on SAXBAG, STREND, and DIFFSTD are denoted as $d'_s(i,j)$, $d'_t(i,j)$, and $d'_f(i,j)$, respectively.

Assignment of Weights to Different Measures The selected measures may not be equally effective in detecting anomalousness; better results can be obtained by assigning the highest weight to that measure which is more effective. The weight associated with measure ℓ of the ith series is computed by finding its k nearest neighbors of the series and the RBDA value is calculated to assign the weight of each measure [60].

Algorithm MUDIM

Require: a positive integer k and a time series dataset $\mathcal{D}$.
Ensure: outliers in $\mathcal{D}$.

Step 1. Calculate the distances between ith and jth time series using SAXBAG, STREND, and DIFFSTD, denoted as $d_s(i,j)$, $d_t(i,j)$, and $d_f(i,j)$, respectively.

Step 2. For $\ell = s,t,f$ normalize the raw distance to be between 0 and 1 as follows:

$$d_\ell(i,j) = \frac{\text{dist}_\ell(i,j)}{mean([5\% \ldots 95\%] \text{ of sorted list of } \text{dist}_\ell(i,j))}$$

Step 3. The weight for the ith series, for the normalized ℓ feature for $\ell \in \{s,t,f\}$, is calculated as follows (cf. Eq. 7.4):

$$w'_\ell(i) = \mathcal{O}_k(i) \times \frac{\sum_{q \in \mathcal{N}_k(i)} d(q,i)}{|\mathcal{N}_k(i)|}$$

Step 4. For all $\mathcal{X}_i \in \mathcal{D}$, find the average distance to the k nearest neighbors of the ith time series using the equation

$$d'_\ell(i) = \frac{\sum_{j \in \mathcal{N}_k(i)} d'_\ell(i,j)}{|\mathcal{N}_k(i)|}$$

where $j = 1, \ldots, m$; $\ell \in \{s,t,f\}$; and $\mathcal{N}_k(i)$ denotes the set of k nearest neighbors of the ith time series.

Step 5. Calculate the anomaly score, $A(\mathcal{X}_i)$, for the ith time series as combined distance based on weighted distances:

$$A(\mathcal{X}_i) = \sqrt{\sum_{\ell \in \{s,t,f\}} (d'_\ell(i))^2 \times w'_\ell(i)}.$$

Step 6. Sort $A(\mathcal{X}_i)$'s in descending order and calculate the average and standard deviation of lower 80% of sorted observations. The series with $A(\mathcal{X}_i)$ larger than (mean + 3× standard deviation) is declared an anomalous.

---

[3] An extreme observation can strongly affect the mean and the standard deviation of a data set. To achieve 'improved' performance and robustness, *trimmed* statistical measures should be computed, after deleting extreme values from the data set.

9.4.2   Identification of Anomalous Series

After obtaining the anomaly score for each series, the final step involves deciding which series is anomalous, i.e., which $O(\mathcal{X}_i)$ value is large enough. This question can be answered using threshold-based or clustering-based approaches:

1. The anomaly score $O(\mathcal{X}_i)$ for each series $\mathcal{X}_i$ is calculated based on distances and weights of the three measures mentioned in the previous section:

$$O(\mathcal{X}_i) = \sqrt{\sum_{\ell \in \{s,t,f\}} (\text{dist}'_\ell(i))^2 \times \text{weight}'_\ell(i)}. \tag{9.1}$$

   To minimize the distortions caused by a few large $O(\mathcal{X}_i)$ values, we may eliminate the highest 20% of the $O(\mathcal{X}_i)$ values, and calculate the (trimmed) mean μ and standard deviation σ of the lower 80% of $O(\mathcal{X}_i)$ values [60]. The ith series is considered anomalous if $O(\mathcal{X}_i) > \mu + 3\sigma$.

2. The second method is to identify potentially anomalous series using a clustering algorithm. This idea can be applied in at least two ways:

   - Apply a clustering algorithm (such as $\mathcal{NC}(\ell, m^*)$) in 3-dimensional space (using all three features of a series). If a series belongs to a cluster of minimal size, then it is declared as normal, otherwise it is considered to be potentially anomalous.[4]
   - Apply the clustering algorithm separately in each feature space. If a series belongs to a cluster of minimal size in each of the three feature spaces, then it is declared as a normal series, otherwise it is potentially anomalous.

   Finally, each potentially anomalous series can be compared with normal series, using the $\mu + 3\sigma$ criterion with the trimmed statistics, as described earlier.

9.5   Online Anomaly Detection for Time Series

As mentioned in Sect. 9.1, online anomaly detection requires rapidly detecting outliers as they arise, where the anomaly is with respect to the most recent data, since we allow for the possibility that the underlying system behavior changes with time. In other words, inferences made from old data may not be valid, and any model used earlier must be retrained, modifying model parameters as needed. The focus of online anomaly detection algorithms is hence on rapidly retraining model parameters, in order to decide about the anomalousness of the most recent observation (or the subsequence ending with the most recent observation), To

---

[4]For example, if there are n time series, then we may consider elements of a cluster containing fewer than n^a time series to be anomalous, where $a = 0.25$.

improve the efficiency of the online algorithm, we may keep a fixed number of observations in the training set, discarding some old data as the current model parameters are updated. Such an approach is sketched in Algorithms "Online model-based subsequence detection" and "Online between-sequence anomaly detection"; the latter applies a specific interpretation of between-sequence anomaly detection, in terms of the relative predictability of elements of a time series from others.

Algorithm Online model-based subsequence detection

1: GIVEN: Time series $\mathscr{X}$, retraining interval w, parametric model M;
2: Train the model M using all available data, in order to predict the next value in time series $\mathscr{X}$;
3: **while** new data continues to arrive, **do**
4:    **if** retraining interval w has passed since the most recent training has occurred, **then**
5:       Update the model M using the most recent w values of time series $\mathscr{X}$ to determine appropriate parameter values;
6:    **end if**
7:    Report the most recent subsequence $x[i], x[i+1], \ldots, x[i+w-1]$ as an anomaly if it is substantially different from the subsequence $x^*[i], x^*[i+1], \ldots, x^*[i+w-1]$ predicted by M;
8: **end while**

Algorithm Online between-sequence anomaly detection

1: GIVEN: Dataset of time series $D = \{X_1, \ldots, X_n\}$, retraining interval w, parametric model M;
2: Train the model M using all available data, in order to predict the next values in each time series in D;
3: **while** new data continues to arrive, **do**
4:    **if** retraining interval w has passed since the most recent training has occurred, **then**
5:       Update the model M using the most recent w values of all time series in D to determine appropriate parameter values;
6:    **end if**
7:    **for** each series $X_j \in D$, **do**
8:       Compute the new predicted subsequence $X_j * [i], X_j * [i+1], \ldots, X_j * [i+w-1]$;
9:       Let $\delta_j[i]$ measure the deviation of this predicted subsequence from the actual observed subsequence $X_j[i], X_j[i+1], \ldots, X_j[i+w-1]$;
10:    **end for**
11:    Let μ_δ and σ_δ be the mean and standard deviation of $\{\delta_1[i], \ldots, \delta_n[i]\}$ values;
12:    Let $\alpha_j = (\delta_j[i] - \mu_\delta)/\sigma_\delta$;
13:    Report $(X_j[i], X_j[i+1], \ldots, X_j[i+w-1])$ as an anomaly if $\alpha_j > 3$.
14: **end while**

9.5.1   Online Updating of Distance Measures

In Sect. 9.4, we discussed an anomaly detection method based on DiffStD, STrend, and SAXBAG measures. Anomaly detection algorithms based on computing distances between time series can be recast as online algorithms, as shown in Algorithm "Online distance-based anomaly detection".

Algorithm Online distance-based anomaly detection

1: GIVEN: Dataset of time series $D = \{X_1, \ldots, X_n\}$, retraining interval w, distance measure d that computes distance between two time series;
2: **while** new data continues to arrive, **do**
3:    Update the distance measure values $d(X_i, X_j)$;
4:    Compute the anomaly scores $\alpha(X_i)$ for each time series, based on the distances, e.g., $\alpha(X_i) = (\delta(X_i) - \mu_\delta)/\sigma_\delta$; where $\delta(X_i)$ = the mean distance from X_i to its nearest k neighbors, μ_δ is an average of this quantity over all time series, and σ_δ is the corresponding standard deviation.
5:    Report any X_j as an anomaly if $\alpha(X_j) > 3$.
6: **end while**

The key is in efficiently updating the distance measure values; for the three distance measures discussed in Sect. 9.4, updates can be performed as follows:

- **DiffStD**: The variance of differences between series $\mathscr{X}_i$ and $\mathscr{X}_j$ at time n can be calculated as:

$$d_f(i,j) = \frac{n \times ssq(i,j) - (sqs(i,j))^2}{n \times (n-1)}, \tag{9.2}$$

where

$$ssq(i,j) = \sum_{t=1}^{n}(x_i(t) - x_j(t))^2 \text{ and } sqs(i,j) = \sum_{t=1}^{n}|x_i(t) - x_j(t)|.$$

The numerator in Eq. (9.2) can be updated for the $(n+1)$th observations by adding $(x_i(n+1) - x_j(n+1))^2$ and $|x_i(n+1) - x_j(n+1)|$ to $ssq(i,j)$ and $sqs(i,j)$ respectively.
- **STrend:** Let $\Delta x_i(n) = x_i(n) - x_i(n-1)$. Then, by definition,

$$S_{i,j}(n) = \begin{cases} 1 & \text{if } \Delta x_i(n) \cdot \Delta x_j(n) > 0 \text{ or } \Delta x_i(n) = \Delta x_j(n) = 0 \\ 0 & \text{otherwise} \end{cases}$$

Consequently,

$$d_t(i,j) = \frac{\sum_{t=2}^{n} S_{i,j}}{n-1}. \tag{9.3}$$

Therefore, to update this value using the $(n+1)$th observation, we modify the numerator by adding the last trend term $S_{i,j}(n+1)$, and accordingly modify the denominator as well.
- **SAXBAG:** This approach converts the data segment in the sliding window of size w to a single SAX word, and then counts the frequency of each word. When data at time $n+1$ is observed, a new SAX word will be generated based on the sequence

$$x_i(n+2-w), x_i(n+3-w), x_i(n+4-w), \ldots, x_i(n+1).$$

The stored data set can be updated to account for the new SAX word.

The updated distances, $d_f(i,j)$, $d_t(i,j)$, and $d_s(i,j)$, are normalized to $d'_f(i,j)$, $d'_t(i,j)$, and $d'_s(i,j)$ respectively, according to Step 2 in Algorithm "MUDIM" and the $w'_\ell(i)$ values are calculated and the anomaly score for the ith series is calculated as follows:

$$A(\mathcal{X}_i) = \sqrt{\sum_{\ell \in \{s,t,f\}} (d'_\ell(i))^2 \times w'_\ell(i)}.$$

Based on the above equations, a "naive" online detection algorithm (NMUDIM) can be developed. Anomaly scores (O_i values) can be plotted for each time series, as illustrated in Fig. 9.10.

Ignoring the length of a time series, the time complexity of NMUDIM is $O(m^2)$, because distances $d_l(i,j)$, $l = s,t,f$ are calculated for all $i \neq j$. In addition, the k nearest neighbors of series i are identified for each i. A method to reduce this complexity is presented in the following paragraph.

Unweighted Multi-Distance Measure (UMUD)   To speed up the online algorithm, a simpler distance measure can be used without considering rank:

$$d_{umd}(i,j) = \sqrt{\frac{\sum_{\ell \in \{s,t,f\}} (d'_\ell(i,j))^2}{3}}, \tag{9.4}$$

and the anomaly score for the ith series, $A_i(k)$, can be defined as the average distance to its k nearest neighbors:

$$A(\mathcal{X}_i) = \frac{\sum_{j \in \mathcal{N}_k(i)} d_{umd}(i,j)}{|\mathcal{N}_k(i)|}. \tag{9.5}$$

Now, select any k neighbors of $\mathcal{X}_i$, and let $\hat{A}(\mathcal{X}_i)$ denote the average $d_{umd}(i,j)$ over them. Then the average distance of k-nearest-neighbors of $\mathcal{X}_i$ must be less than or equal to the average distance of any k neighbors of $\mathcal{X}_i$, so:

$$A(\mathcal{X}_i) \leq \hat{A}(\mathcal{X}_i).$$

In addition, if we can find a threshold λ such that $A)\mathcal{X}_i) < \lambda$ implies $\mathcal{X}_i$ is not an anomalous series; then any $\hat{A}(\mathcal{X}_j) < \lambda$ also implies $\mathcal{X}_j$ is not an anomalous series either; thus most of the non-anomalous series can be excluded from anomaly score calculations. To find an estimate of the threshold, λ, the following sampling procedure is applied: calculate $A(\mathcal{X}_i)$'s values for $i \in S$, where S contains a small fraction (α) of the elements in $\mathcal{D}$, randomly selected. Then λ is chosen to equal the value of $A(\mathcal{X}_i)$ which is at the top ($\beta \times 100$)th percentile in descending order. Based on the above observations, a faster version of MUDIM is presented in Algorithm "MUASD", whose key steps are as follows, repeated as new data points arrive:

1. Find λ as described above.
2. For $\mathcal{X}_i \in \mathcal{D} - \mathcal{S}$, maintain a binary max heap consisting of $dist_{msm}(i, j)$ where various $\mathcal{X}_j$'s are selected k neighbors of $\mathcal{X}_i$. If the average of these k neighbors is less than λ, then $\mathcal{X}_i$ is declared as non-anomalous. Else $dist_{msm}(i, j)$ is calculated for next selected value of j, and the heap is updated by keeping only the smallest k values of $dist_{msm}$. The anomalousness of series $\mathcal{X}_i$ is tested using the above criterion. This process stops if at any stage, the series is found to be non-anomalous or none of the k neighbors remain.
3. Calculate the anomaly scores of all potential anomalous series (found in Step 2) and identify the anomalous series, if any.

Algorithm MUASD Online detection of time series anomalies

Require: Set $\mathcal{D}$ of m time series $\mathcal{X}_i$, each initially of length n, growing as new $\mathcal{X}_i(t)$ values arrive;
Ensure: Calculate and store all pairwise distances between time series in $\mathcal{D}$ upto length n.
1: **while** New data values arrive over time, **do**
2:     Update and store pairwise (sequence-sequence) distance values using efficient heap data structures;
3:     **for each** $\mathcal{X}_i$, **do**
4:         update aggregated distances $\mathbf{a}_i$ from $\mathcal{X}_i$ to the nearest neighbors of $\mathcal{X}_i$
5:         **if** $\mathbf{a}_i$ values are small enough, **then**
6:             mark $\mathcal{X}_i$ as non-anomalous;
7:         **else**
8:             compute the current anomaly score of $\mathcal{X}_i$ from $\mathbf{a}_i$;
9:         **end if**
10:    **end for**
11: **end while**

By applying these techniques, the time complexity for the online algorithm is considerably reduced, as verified in experimental simulations whose results are reported in Sect. 9.6.

9.5.2 Multiple Measure Based Abnormal Subsequence Detection Algorithm (MUASD)

An example problem is shown in Fig. 9.6, where each subsequence represents power consumption on a weekly basis. Such anomalies are detectable if the time period over which each subsequence is examined spans one week, whereas these anomalies are not detectable if the time span is a day or a month. Hence a critical parameter is the window (w) or length of the subsequences for which anomalousness needs to be

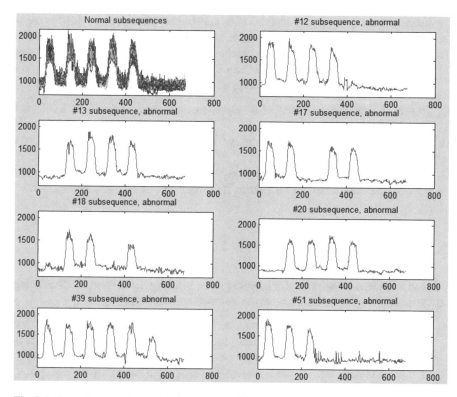

Fig. 9.6 Examples for abnormal subsequences: Subsequences extracted from a long series of power consumption for a year; each subsequence represents power consumption for a week. (*Top left*) contains normal subsequences; others contains abnormal subsequences along with the week they appear. Abnormal subsequences typically represent the power consumption during a week with special events or holidays

calculated, and this may be a user-provided parameter; typically w is much smaller than the length n of the entire time series.

Detection of an abnormal subsequence can be recast as the problem of comparing each subsequence of a given length w to other subsequences (also of length w) in the time series of length n, which is similar to the problem considered earlier in this chapter, e.g., using MUDIM algorithm. More precisely: given a time series X, the set of extracted subsequences, $X_w = \{X_{p,w}; 1 \le p \le (n - w + 1)\}$, consists of $X_{1,w} = \{x(1), x(2), \ldots, x(w)\}$; $X_{2,w} = \{x(2), x(3), \ldots, x(w + 1)\}$; $\ldots$; $X_{n-w+1,w} = \{x(n-w+1), x(n-w+2), \ldots, x(n)\}$. We must determine if any $X_{i,w}$ is substantially different from the others, e.g., by evaluating whether its average distance to its k nearest neighbors is much larger than is the average distance for other subsequences.

We expect that the subsequence $X_{i,w}$ is substantially similar to the subsequence $X_{i+1,w}$, since consecutive series share most elements, and are likely to be nearest neighbors with only a small distance between them. To address this *self-match* problem, a subsequence is compared with another subsequence provided there is no overlap between the two subsequences. Now an algorithm such as MUDIM can be applied to the set of series in X_w, finding the k nearest neighbor subsequences that do not overlap with each subsequence $X_{p,w} \in X_w$.

The following steps can improve computational efficiency:

1. To find the nearest neighbor of a subsequence in X_w, we may use Euclidean distance.
2. Instead of multiple nearest neighbors, we may use only the nearest neighbor, i.e., $k = 1$.
3. Frequencies of SAX words of a subsequence $X_{p,w}$ may be compared with their frequencies over the entire time series X (instead of other subsequences).

9.5.3 Finding Nearest Neighbor by Early Abandoning

To search the nearest neighbor for a subsequence in a long time series, a "Reordering early abandoning" approach can reduce computational effort [97].

- In this approach, a promising candidate is first chosen as the nearest neighbor for a subsequence $X_{i,w}$, and the least among all distances calculated so far is called its best-so-far distance. Suppose the nearest neighbor for $X_{1,w}$ is $X_{p,w}$, then what is the possible nearest neighbor for $X_{2,w}$? Since most objects in $X_{1,w}$ and $X_{2,w}$ are identical, then $X_{p+1,w}$ would have a very high possibility to be the nearest neighbor for $X_{2,w}$. Thus, instead of searching nearest neighbor from the beginning, we can start at $X_{p+1,w}$, and its distance to $X_{2,w}$ is the initial best-so-far, as we begin examining other possibilities.
- At each subsequent step during distance calculations with another subsequence to evaluate whether it could be a nearest neighbor of $X_{i,w}$, computation can be terminated if the current accumulated distance exceeds best-so-far distance, and we can abandon this candidate since it cannot be the nearest neighbor for $X_{i,w}$. Performing such calculations beginning with large distance values (instead of just going from left to right) will result in quick termination, as in the case of other variants of the *Branch-and-bound* approach.

Figure 9.7 depicts an example of this approach, when the current value of *best-so-far* exceeds the sum of the distances for the five successive (nearest) time points, but not the sum of the three largest distance values among the set of time points to be considered, when comparing time series T1 with time series T2. Without the reordering step, six distance additions are required (left part of the figure), but three are sufficient (right part of the figure) when the largest distances are considered first.

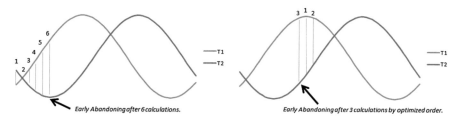

Fig. 9.7 *Left* (No reordering): 6 calculations performed before abandoning; *Right* (With reordering): Abandoning after 3 largest distance computations

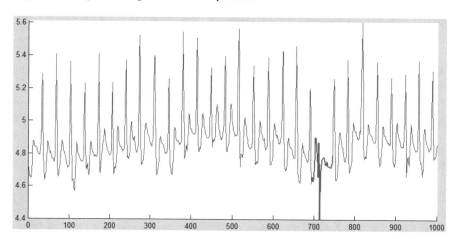

Fig. 9.8 Sub-sequence frequency—*Red bold line* represents a subsequence considered abnormal since it appears only once in this series, whereas other subsequence patterns occur much more frequently

9.5.4   Finding Abnormal Subsequence Based on Ratio of Frequencies (SAXFR)

Another approach, called *SAX words based frequency ratio (SAXFR),* is based on assuming that the occurrence frequencies of abnormal subsequences are far lower than the frequencies of normal subsequences. An example is shown in Fig. 9.8. To implement SAXFR we calculate the ratios between frequencies of SAX words generated from an abnormal subsequence with frequencies of these words in the entire series. The ratio computed for an abnormal subsequence is expected to be much higher than the ratio computed for a normal series. The anomalousness measure is then defined to be $\alpha_{SAXFR}(X)$ = average of the reciprocals of the frequencies of SAX words contained in X.

For example, if three SAX words abc, aac, abd are generated from an abnormal subsequence, and these words appear $2, 3, 2$ times respectively within the entire series, then the corresponding reciprocals are $1/2, 1/3, 1/2$. The average value can be calculated for such reciprocal frequencies, in this case, it is $\alpha_{SAXFR}(X) = (0.50 + 0.33 + 0.50)/3 = 0.44$. For comparison, consider another subsequence, each of whose SAX words appear 10 times; then the corresponding average value would instead be 0.1, a much lower value, indicating that this latter subsequence is not anomalous.

In capturing the outlierness of subsequences, the window size plays a critical role. If the window size of each word in SAX is too short, then the shorter subsequence represented by each SAX word may not be anomalous and we have too many false negatives. If the window size is too long, then the number of SAX words obtained in each subsequence (of length w) is less, and their frequencies may be low even if the sequence is not anomalous, thereby impacting the results of the SAXFR approach in a different way.

In SAXFR since the ratio is compared between a subsequence and the entire series, there is no need for nearest neighbor computation, and there is also no need for additional parameters. Another advantage is that frequencies can be precomputed over the entire sequence, and subsequence SAX frequencies can be updated incrementally, so the computation is very fast.

9.5.4.1 Effect of SAXFR Subsequence Length Parameter

Figure 9.9 summarizes the results of some experiments to identify the relationship between subsequence length and performance of SAXFR. Other parameters had the following values in these simulations: the sliding window size is $w/2$, the number of symbols is 4, and the length of the SAX word is 5. Three data sets (SYN0, ECG1 and ECG2) were used with subsequence lengths in $\{10, 20, \ldots, 120\}$. The

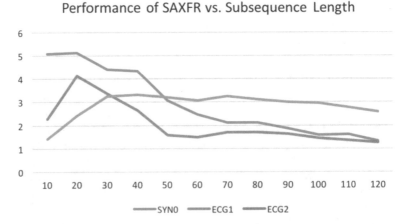

Fig. 9.9 Performance of SAXFR versus length of the subsequence. In general, the ratio of anomaly scores for abnormal subsequence to normal subsequence decreases with the size of the subsequence

minimum sizes of abnormal subsequences in these three data sets were 40, 40 and 100 respectively. In all the runs of the algorithms performed, the average anomaly scores of abnormal subsequences were found to be significantly higher than that of normal subsequences.

In general, the size of a subsequence does not have to be necessarily equal to the exact size of the abnormal subsequence in order to achieve the best performance. In SYN0, α_{SAXFR} of each abnormal is three times larger than that of a normal subsequence, when the subsequence length is between 10 and 50. When the subsequence length increases, the α_{SAXFR} value decreases. In ECG1, the ratio between the α_{SAXFR} values of abnormal vs. normal subsequences exceeds 3 when the subsequence length is 20 or 30. In ECG2, the ratio is over 3 when the subsequence length is between 30 and 80.

Experimental results suggest that too small or too large subsequence lengths are not the best choices for SAXFR. If the size is too small, then the SAX word extracted from an abnormal subsequence may be similar to a SAX word extracted from normal subsequences. On the other hand, if the size is too large, then the ratio remains small, causing a similar problem. The best window size appears to be data-dependent: one possible heuristic is to use any known periodicity in the data set to determine the window length.

The other factor that impacts the performance is the degree of similarity between normal subsequences. If normal subsequences are highly similar to each other, as in the SYN0 data, then the ratio of average anomaly score of abnormal subsequence over that of normal subsequence is larger.

9.5.5   MUASD Algorithm

The above discussion leads to a simple definition of the steps for the MUASD algorithm:

1. Given a time series X, length of subsequence w, obtain X_w, the set of all possible subsequences of X using sliding window technique.
2. Calculate $dist_s$, the SAX word distance between s_i and its nearest neighbor.
3. Calculate the frequencies ($FreqAll$) of all SAX words based on the entire series.
4. For each subsequence $X_{p,w} \in X_w$, calculate the frequency $Freq_i$ for $X_{p,w}$, and the associated distance measure $dist_r$.
5. Calculate the same-trend distance $dist_t(i)$ between s_i and its nearest neighbor.
6. Normalize $dist_s, dist_r, dist_t$ to compute the z-scores (subtracting the mean and dividing by the standard deviation).
7. Combine all distances and compute an anomaly score $O(i)$ for each ith subsequence.

In this algorithm, different weights are not assigned to different features (in contrast to MUDIM). The reason for this is that the rank based weight allocation method works better if the number of outliers is far less than the number of normal objects. But in the abnormal subsequence detection problem, the number of objects contained in an abnormal subsequences is sometimes large, thus the number of subsequences containing common data objects with abnormal subsequence is also large. For instance, in the ECG1 data set, the total number of data objects with abnormal subsequences is close to 10% of the entire series. A large number of anomalies will bias the results of rank based methods.

Algorithm MUASD Algorithm to calculate anomaly scores for each subsequence

---

Require: Time series $\mathscr{X}$ of length n, subsequence length parameter $w \ll n$;
Ensure: anomaly scores for subsequences;
1: **for** Each $i \in \{1, \ldots, n - w + 1\}$, **do**
2:    Let X_i be the result (with mean 0) of normalizing the subsequence $[\mathscr{X}[i], \ldots, \mathscr{X}[i+w-1]]$;
3:    Aggregate the three distances ($dist_s, dist_t, dist_r$) from X_i with each X_j, for $j < i$;
4: **end for**
5: **for** Each $i \in \{1, \ldots, n - w + 1\}$, **do**
6:    For X_i, find the k nearest subsequences using the aggregated distances;
7:    Compute the anomaly score of X_i, combining the distances to its k nearest neighbors;
8: **end for**

9.6   Experimental Results

This section compares the performance of algorithms discussed in the previous sections for all three problems. These algorithms include distance-based algorithms using various measures as well as algorithms based on the combination of measures discussed in Sect. 9.4.

9.6.1   Detection of Anomalous Series in a Dataset

The results described below use the *RankPower* comparison metric [103], described in Sect. 2.1.1. According to this criterion, better algorithms exhibit higher RankPower values; i.e., the RankPower of an algorithm is 1 (the largest possible value) if all the k known outlier series in $\mathscr{D}$ are ranked between 1 and k by the algorithm.

Datasets We have used 47 datasets, consisting of three synthetic datasets and 44 modified real datasets from multiple application domains, to compare the performance of each algorithm, including MUDIM. In these experiments, parameters of SAXBAG, *viz.*, the size of subsequence, the length of SAX word and the number of symbols, were 20, 5, and 5, respectively. All time series in the data sets are first normalized to have zero mean and standard deviation equal to one [71].

Synthetic datasets have been designed to introduce typical time series problems such as time lagging and amplitude differences (see Fig. B.1). The real datasets come from different application domains, such as finance, electronics, image recognition and video surveillance. We augment these real datasets by introducing one or more anomalous series. Some background information about the data sets is given in Sect. B.1.2.

Some data sets, originally designed for classification problems, were modified for anomaly detection experiments by selecting all time series from one class and choosing a few time series from another class that are very different from most of the series of the first class. For example, 'Commodity Prices' dataset contains five commodity prices from [46] and one consumer retailer stock [118]; an 'outlier' commodity price time series.

Details about the normal series and anomalous series in each dataset are given in Tables B.2 and B.3.

Results RankPowers of all algorithms on all data sets are summarized in Table 9.2, with the best results shown in boldface. From this table, we conclude that MUDIM has the best overall performance and its RankPower values are 1 in all cases; i.e., it finds the anomalous series in all data sets; confirming that a combination of measures performs better in most cases. Simpler methods such as DIFFSTD, DTW, SAXSL, and FOURIER also work well in some, but not all cases. Domain specific analysis is summarized below:

- MUDIM and STREND are good at detecting outliers in financial data sets such as stock prices and commodity prices, perhaps because financial time series can be easily aligned, and variation in amplitudes is a common problem in financial time series (for example, one stock may increase by 1% and another stock may increase by 5% in the same day). Measures such as SAXSL, SAXBAG, DIFFSTD and FOURIER rarely discover real outliers in such data.
- Traditional methods, such as DTW, WAVELET and FOURIER, show good performance with many datasets.

9.6.2 Online Anomaly Detection

Datasets The algorithms are evaluated using three synthetic data sets and eight modified real data sets introduced earlier in Sect. 9.6.1, viz., SYN1, SYN2, SYN3, STOCKS, OPMZ, MOTOR, POWER, TEK140, TEK160, TEK170 and SHAPE1. Key characteristics of the data sets are shown in Tables B.2 and B.3.

In all experiments, the initialization is performed at $l = 100$th observations and the rest of the observations of the series are used to test the effectiveness of the proposed algorithm. The number of nearest neighbors is set at $k = 5$ for all data sets.

Table 9.2 RankPower comparison of algorithms; using 47 data sets

	SAXSL	SAXNW	SAXBAG	DTW	DIFFSTD	STREND	WAVELET	FOURIER	MUDIM
Average	0.733	0.714	0.679	0.722	0.686	0.587	0.685	0.711	**0.922**
Relative ranking based on Avg. RankPower	2	4	8	3	6	9	7	5	1
Number of datasets on which perfect results were achieved	25	26	25	27	23	18	24	25	**40**
Number of datasets on which RankPower was at least 0.5	37	36	31	33	34	29	32	35	**45**
Relative ranking based on No. of datasets with RankPower ≥ 0.5	2	3	8	6	5	9	7	4	1

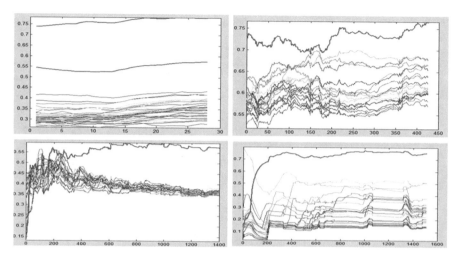

Fig. 9.10 NMUDIM anomaly scores for each data set. Plots of the online anomaly scores for each series at time 100+x in four data sets (*from top to bottom, left to right*) : SYN2, STOCKS, MOTOR, and SHAPE1. *Red curves* indicate anomalous series

Results The time series in the data sets are plotted in Fig. B.4 and the performance of the NMUDIM algorithm is shown in Fig. 9.10. As more data arrives and if more unusual patterns occur, the anomaly score increases and the gap between anomalous and normal series becomes larger. Normal series' anomaly scores converge if they are similar to each other.

The above algorithms are compared with three other online detection algorithms based on (a) Euclidean distance, (b) Dynamic Time Warping (DTW), and (c) Autoregressive (AR) approach, proposed by [11, 72] and [45] respectively. The first two of these methods calculate a measure of anomalousness of a time series by (i) finding the k nearest neighbors of the series, and (ii) using the average distance of these k neighbors. The third method constructs a global AR model for all series and then measures the anomaly score at time t as the gap between the observed value and the predicted value.

We compare the numbers of true anomalous series detected by these algorithms after all observations have arrived, as shown in Table 9.3. It can be seen that NMUDIM and UMUD perform very well for all 11 data sets; i.e., anomalous series are always in top p places. Other methods do well for some but not all sets. This illustrates that the use of multiple criteria is important in order to capture multiple types of anomalies.

Table 9.3 Performance of all algorithms; numbers show the true outliers that are identified by algorithms

Data sets	p =# of true outliers	Euclid	AR	DTW	NMUDIM	UMUD
SYN1	2	2	1	2	2	2
SYN2	2	2	2	2	2	2
SYN3	1	0	1	0	1	1
STOCKS	1	0	0	0	1	1
OPMZ	1	0	1	0	1	1
MOTOR	1	1	0	1	1	1
POWER	7	7	1	7	7	7
TEK140	1	1	0	1	1	1
TEK160	1	1	0	1	1	1
TEK170	1	1	0	1	1	1
SHAPE1	1	1	0	1	1	1

9.6.3   Anomalous Subsequence Detection

We compare the performance of several algorithms as described below:

- (Euclidean distance method) Keogh et al. [72] use the Euclidean distance to the nearest neighbor as anomaly score to find the most unusual subsequence. We apply their method for every subsequence so that each subsequence can be assigned an anomaly score.
- (SAX Frequency Ratio) We also applied SAXFR separately to check its performance. The size of feature window is set to $w/2$. Alphabet size is set to 4 as suggest by Rakthanmanon et al. [97] and word size is set to 5. The algorithm parameters are chosen for the best performance.
- Model prediction based methods have been used, e.g., Auto Regression by Fujimaki et al. [45], ARMA by Pincombe [94], ARIMA by Moayedi et al. [87]. Models are generated based on the entire series, and then the model is applied to predict the next observation using the model. The mean squared error is considered to be the anomaly score. Parameter values are chosen from 10 to 20, yielding the minimum mean squared error.

To evaluate each algorithm, we compare the anomaly scores of abnormal subsequences (which are known to us) with those of normal subsequences. If the anomaly scores of the abnormal subsequences are higher than those of the normal subsequences, then the algorithm is considered to be effective; otherwise not.

Data Sets Results are shown for one synthetic data set and 7 real data sets. The synthetic data set contains several copies of 'normal' data and two identical abnormal subsequences. Practical data sets are obtained from multiple application domains: two data sets are from medical domain, three are from electronic domain and two are from video surveillance domain. The details of data sets are given in

Table B.4. The length of *w* for each data set is chosen mostly by following the original author's suggestion.

Results   Experimental results are shown in Fig. 9.11; results for other datasets are in Appendix B.1.5.

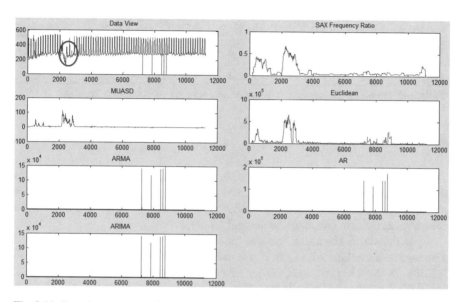

Fig. 9.11 Experimental results for VIDEOS1. *Red circle* highlights abnormal subsequences. (*Top Left*) Plot of VIDEOS1 time series; (Other) Results of 6 algorithms used in these comparisons. Y-axis represents anomaly scores at time *t*. X-axis shows time *t*

Our observations are:

- Euclidean distance based method works well if abnormal subsequences are not similar to each other, see example in Fig. B.5. This approach also results in false positives for some data sets such as TEK140 and TEK170.
- SAXFR method works well if normal subsequences appear more frequently. In TEK data sets, SAXFR doesn't work well since normal subsequences only appear four more times than abnormal subsequences and contain considerable noise.
- Model based methods appear to be satisfactory for data set ECG1 in Fig. B.6, but only one of three abnormal subsequences is detected. For other data sets, the model based methods did not succeed in finding anomalies.
- MUASD usually works better for high frequency data sets. In low frequency data sets such as TEK, it may result in false positives.

Table 9.4 Running time of NMUDIM and average computation workload comparison between NMUDIM and UMUD

Running Time	Length	# of Series	NMUDIM (Seconds)	NMUDIM Workload	UMUD Workload	Ratio
Synthetic1	1500	20	3.90	190	63	33.2%
Synthetic2	1500	40	8.20	780	405	51.9%
Synthetic3	1500	80	18.28	3160	1138	36.0%
Synthetic4	1500	200	53.60	19900	6535	32.8%
Synthetic5	1500	400	152.69	79800	28387	35.6%
Synthetic6	1500	1000	706.64	499500	113304	22.7%

"Workload" represents the average number of comparisons performed in each iteration.

9.6.4 Computational Effort

The time complexity of the UMUD algorithm is $O(n \times m^2)$ because it updates stored data structures when new data arrive and then inter-time series distances are obtained for each pair of series. In addition, the k nearest neighbors need to be computed in order to calculate the anomaly scores.

Table 9.4 shows the computational times (in seconds) needed for six synthetic data sets, on a machine with Core i7, 6G memory, Windows 7 system, using Matlab R2010b. In those experiments, the parameters for UMUD were as follows: k is 1, α is 0.05 if number of series is less than 200, otherwise 0.1. β is 0.1. UMUD was about 60% faster than NMUDIM. We find that the anomalous series begins to differ from the rest of the group within as few as 100 additional observations.

The comparisons of computational effort (time) between the NMUDIM and UMUD are shown in Table 9.4. The MUDIM approach is efficient and detects anomalous series as soon as it begins to drift away from the other (non-anomalous) series, a substantial advantage over other anomaly detection algorithms for time series. This approach can handle data uncertainty very well, and its online version does not require any training data sets. Compared with other methods, it requires less domain knowledge.

9.7 Conclusion

Many applications require the analysis of data that arrives over time, identifying anomalies that occur within a sequence, as well as between sequences, possibly requiring rapid identification of the anomalies as soon as new data arrive. As in non-time-dependent anomaly detection, critical questions concern the choice of a distance measure and the possible transformation of the data into an appropriate space that facilitates anomaly identification. Additional complexity enters the analysis task due to the problem-dependent nature of appropriate values for critical parameters such as the appropriate time window sizes or the characteristics of the anomalies of interest.

In this chapter, we have described and evaluated the performance of popularly used measures and algorithms for anomaly detection with time series data. In particular, approaches that involve discretization of the time series data appear to be useful in many applications, along with distance measures specialized for various transformations. Comparisons across multiple time series must account for the possibility of time lags and lack of synchronicity between different time series. Comparisons between subsequences of the same time series must recognize that each subsequence needs to be evaluated within the context of surrounding values.

Algorithms that focus on a single distance measure (or transformation) often miss anomalies that are adequately captured by other measures or transformations. Combinations of multiple measures hence perform best, as substantiated by empirical results that showed that a combination of three measures performed the best, i.e., was able to identify anomalous times series in all domains considered in various simulations. The selection of three measures is based on some preliminary analyses, with measures that do not overlap in their detection capabilities. For example, the MUASD algorithm combines three important processes: early abandoning searching for the nearest neighbor subsequence, frequencies ratio comparisons for each subsequence, and calculation for the same trend. This approach performs best on between-sequence as well as within-sequence anomaly detection, although the time complexity is higher than that of algorithms that focus on a single distance measure.

Appendix A
Datasets for Evaluation

The datasets used in evaluating the performances of anomaly detection algorithms are summarized below.

A.1 Synthetic Datasets

Two synthetic datasets, shown in Figs. A.1 and A.2, are used to evaluate the outlier detection algorithms. For illustrative convenience, we use only 2-dimensional data points so that outliers can be seen easily. In each dataset, there are multiple clusters with different densities. In each dataset, we have placed six additional objects, (A, B, C, D, E, and F) in the vicinities of the clusters.

- **Synthetic Dataset 1:** Synthetic dataset 1 contains four clusters of different densities consisting of 36, 8, 8, and 16 instances.
- **Synthetic Dataset 2:** Synthetic dataset 2 consists of 515 data objects including six planted outliers; this data set has one large normally-distributed cluster and two small uniform clusters.

A.2 Real Datasets

We have used three well known datasets, namely the Iris, Ionosphere, and Wisconsin breast cancer datasets. We use two ways to evaluate the effectiveness and accuracy of outlier detection algorithms;

© Springer International Publishing AG 2017
K.G. Mehrotra et al., *Anomaly Detection Principles and Algorithms*, Terrorism, Security, and Computation, https://doi.org/10.1007/978-3-319-67526-8

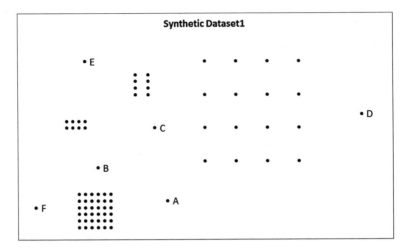

Fig. A.1 A synthetic dataset with clusters obtained by placing all points uniformly with varying degrees of densities

Fig. A.2 A synthetic data set with one cluster obtained using the Gaussian distribution and other clusters by placing points uniformly

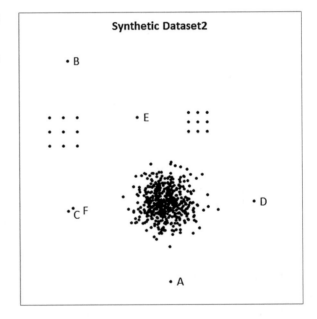

1. Detect rare classes within the datasets. This methodology has also been used by many researchers such as Feng et al., Cao [17], and Tang et al. [42, 107]. Rare classes are generated by removing a majority of objects from the original class; remaining points of this class are considered anomalous points.
2. Plant outliers into the real datasets (according to problem specific knowledge) and expect outlier detection algorithms to identify them.

Iris Dataset This well-known data set contains the categorization of iris flowers to three classes: Iris Setosa, Iris Versicolour, Iris Virginica, with 50 instances each. The Iris Setosa class is linearly separable from the other two classes, but the other two classes are not linearly separable from each other.

- Objects belonging to the Iris Setosa class are made 'rare' (anomalous) by randomly removing 45 instances; the remaining 105 instances are used in the final dataset.
- Three outliers are planted; the first outlier has maximum attribute values, second outlier has minimum attribute values, and the third has two attributes with maximum values and the other two with minimum values.

Johns Hopkins University Ionosphere Dataset The Johns Hopkins University Ionosphere dataset contains 351 data objects with 34 attributes; all attributes are normalized in the range of 0 and 1. There are two classes labeled as good and bad with 225 and 126 data objects respectively. There are no duplicate data objects in the dataset.

- To form the rare class, 116 data objects from the bad class are randomly removed. The final dataset has only 235 data objects with 225 good and 10 bad data objects.
- Both classes, labeled as good and bad, with 225 and 126 instances respectively, are kept in the resulting dataset. Three outliers are inserted into the dataset; the first two outliers have maximum or minimum value in every attribute, and the third has 9 attributes with unexpected values and 25 attributes with maximum or minimum values. The unexpected value here is a value that is valid (between minimum and maximum) but is never observed in the real datasets.[1]

Wisconsin Diagnostic Breast Cancer Dataset[2] Wisconsin diagnostic breast cancer dataset contains 699 instances with 9 attributes. There are many duplicate instances and instances with missing attribute values. After removing all duplicate instances and instances with missing attribute values, 213 instances labeled as benign and 236 instances as malignant were left.

- All 213 instances of the benign class are retained, but 226 malignant instances are randomly removed, leaving 10 'anomalous' points; the final dataset consisted 213 benign instances and 10 malignant instances.
- After removal of duplicate instances and instances with missing attribute values from the original dataset, only 449 instances were kept consisting of 213 instances labeled as benign and 236 as malignant. Next, two outliers were planted into dataset; both outliers have maximum or minimum values for all attributes.

---

[1] For example, one attribute may have a range from 0 to 100, but the value of 12 never appears in real dataset.

[2] http://archive.ics.uci.edu/ml/machine-learning-databases/breast-cancer-wisconsin/.

Basketball Dataset (NBA)[3]   This dataset contains Basketball player statistics from 1951–2009 with 17 features. It contains records of all players' statistics in regular seasons, and another set of records that contain information about all star players for each year. Regular season statistics for all star players (for 2009) are considered as the outliers, and all the other players statistics (also for 2009) are considered as normal points in the dataset.

Smartphone-Based Recognition of Human Activities and Postural Transitions Data Set (ACT)[4]   This dataset consists of data from 30 volunteers from age 19–48; there are three static postures (standing, sitting, lying) and three dynamic activities (walking, walking downstairs and walking upstairs), as well as postural transitions that occurred between the static postures (stand-to-sit, sit-to-stand, sit-to-lie, lie-to-sit, stand-to-lie, and lie-to-stand). Volunteers wore smartphones and data was captured from the 3-axial linear acceleration values and 3-axial angular velocity values from the sensors, with 561 features and 12 classes. The class with the least number of instances (sit-to-stand) is considered to consist of outliers, and the class with most number of instances (standing) is considered to consist of inliers.

A.3   KDD and PED

In addition two more datasets described below are also used.

- **Packed Executables dataset**[5] Executable packing is the most common technique used by computer virus writer to obfuscate malicious code and evade detection by anti-virus software. This dataset was originally collected from the Malfease Project dataset and is used to classify the non-packed executables from packed executables so that only packed executables could be sent to an universal unpacker. In our experiments, we select 1000 packed executables as normal points, insert 8 non-packed executables as anomalies. All the 8 features are used in experiments (http://roberto.perdisci.com/projects/cpexe).
- **KDD 99 dataset**[6] KDD 99 dataset is constructed from DARPA intrusion dataset evaluation program. KDD 99 has been widely used in both intrusion detection and anomaly detection research area. There are 4 main attack categories and normal connections in KDD 99 dataset. In our experiments, we select 1000 normal connections from testing dataset and insert 14 attack connection as anomalies. All the 41 features are used in experiments (http://kdd.ics.uci.edu/databases/kddcup99/kddcup99.html).

---

[3]http://databasebasketball.com/.

[4]http://archive.ics.uci.edu/ml/datasets/Smartphone-Based+Recognition+of+Human\+Activities+and+Postural+Transitions.

[5]http://roberto.perdisci.com/projects/cpexe.

[6]http://kdd.ics.uci.edu/databases/kddcup99/kddcup99.html.

Appendix B
Datasets for Time Series Experiments

B.1 Datasets

We have used 47 data sets, consisting of three synthetic datasets and 44 modified real datasets from variety of application domains. A brief description of these datasets is provided in the following sections and summary in Tables B.2 and B.3.The real datasets come from different application domains, such as synthetic, finance, electronic, image recognition and video surveillance. We augment these real datasets by introducing one or more anomalous series.

B.1.1 Synthetic Datasets

Synthetic datasets are designed to introduce typical time series problems such as time lagging and amplitude differences (see Fig. B.1).

B.1.2 Brief Description of Datasets

Some background information about the datasets is given below (Figs. B.2 and B.3):

- **Synthetic Dataset 1(SYN1)**. Synthetic dataset 1 contains 14 univariate time series including two anomalous time series. The length of each time series is 500. The two anomalous time series have shapes considerably different from the others.

© Springer International Publishing AG 2017
K.G. Mehrotra et al., *Anomaly Detection Principles and Algorithms*, Terrorism, Security, and Computation, https://doi.org/10.1007/978-3-319-67526-8

Table B.1 Time series $\mathscr{X}$ used in SXBAG evauation

0.00	0.31	1.27	0.34	−0.29	−1.08	−0.44	−0.98	−0.57	−0.04
−1.00	−1.96	−1.39	−2.03	−2.84	−3.05	−2.51	−1.90	−1.21	−0.53
−1.04	−1.18	−0.96	−1.06	−1.92	−1.57	−1.46	−0.96	−1.57	−2.34
−1.78	−1.35	−1.85	−1.04	−1.44	−1.20	−0.26	−0.93	−1.02	−1.68
−0.91	−1.83	−2.32	−1.38	−1.46	−1.25	−1.32	−0.73	−1.38	−0.70
−1.00	−1.29	−0.99	−0.52	−0.84	−1.24	−0.36	0.51	1.35	1.41
0.85	−0.12	0.85	−0.04	0.46	0.42	−0.04	−0.09	0.45	0.99
0.91	1.10	1.48	1.67	1.14	0.61	0.13	−0.30	−0.05	−0.20
0.74	0.20	0.69	−0.14	−0.06	−0.90	−0.36	−0.30	−0.19	−0.12
0.17	−0.38	0.34	0.17	0.16	1.14	0.49	−0.13	0.48	0.16
0.20	−0.37	−0.58	−0.99	−1.92	−0.98	−0.51	−0.75	−0.48	−0.49
−0.44	−1.37	−2.01	−2.03	−1.42	−2.04	−2.56	−3.11	−3.86	−3.57

- **Synthetic Dataset 2(SYN2).** Synthetic dataset 2 contains 30 time series including two anomalous time series, each of length 128. The normal time series consist of two dissimilar groups, but the two anomalous series do not belong to either group.
- **Synthetic Dataset 3(SYN3).** Synthetic dataset 3 contains 7 time series including one anomalous time series with the length of 500. The dataset contains time series with many types of scaling such as increasing scaling and varying scaling. The anomalous time series is a single (flat) line perturbed by random noise.

B.1.2.1   Real Datasets

The real datasets come from different application domains, such as finance, electronic, image recognition and video surveillance. Since some data sets were originally designed for classification problems, we have modified them for anomaly detection experiments by selecting all time series from one class and choosing few time series from another class that are very different from the most of the series of the first class, thus making them anomalous.

- **Stocks (STOCKS).** This dataset consists of closing prices of 17 oil & gas operation industry stocks and one consumer retailer stock (WMT:Wal-Mart) from January-4th-2010 to February-10th-2012 [118]. All stock prices are aligned by dates and contain 527 values. The symbols for the 17 stocks in oil and gas industry are APA, APC, BP, CNQ, COP, CVE, CVX, DVN, EOG, HES, IMO, MRO, OXY, STO ,TOT, WNR, XOM. All stock prices are normalized with a mean of zero and standard deviation of one.
- **Commodity Prices (OPMZ).** This data set contains five commodity prices from [46] and one consumer retailer stock(WMT:Wal-Mart) from October-13th-2009 to April-13th-2012 [118]. Each series contains 629 values. The five commodities are wheat, corn, cocoa, coffee and cotton. All prices are normalized with a mean

Table B.2 Summary of all time series sets. Similarity of normal series represents that how similar normal series look like to each other. N represents number of series and O number of outliers

Dataset	Domain	Length	N	O	Normal Series			Anomalous Series		Dissimilarity to normal series			
					Lagging	Amplitude	Similarity	Lagging	Amplitude	<10%	<30%	<60%	≥60%
SYN1	Synthetic	501	14	2	Yes		High						Yes
SYN2	Synthetic	128	30	2	Yes	Yes	High		Yes			Yes	
SYN3	Synthetic	500	7	1		Yes	High						Yes
STOCKS	Finance	527	18	1		Yes	Medium		Yes			Yes	
OPMZ	Finance	629	6	1		Yes	Low		Yes				Yes
EMP3	Astronomic	201	11	3	Yes		High		Yes			Yes	
MOTOR	Electronic	1500	21	1	Yes		High		Yes		Yes		
POWER	Power	672	51	7		Yes	Medium		Yes		Yes		
TEK140	Electronic	1000	5	1	Yes		Medium	Yes	Yes	Yes			
TEK160	Electronic	1000	5	1	Yes		Medium	Yes	Yes	Yes			
TEK170	Electronic	1000	5	1	Yes		Medium	Yes	Yes	Yes			
SHAPE1	Image	1614	21	1	Yes		Medium	Yes	Yes		Yes		
SHAPE2	Image	1614	21	1	Yes		Medium	Yes	Yes		Yes		

Table B.2 (continued)

Dataset	Domain	Length	N	O	Normal Series			Anomalous Series		Dissimilarity to normal series			
					Lagging	Amplitude	Similarity	Lagging	Amplitude	<10%	<30%	<60%	≥60%
ADIAC1	Image	2801	7	1		Yes	Medium		Yes				Yes
ADIAC2	Image	176	43	1			Exact		Yes	Yes			
BEEF1	Unknown	470	7	1		Yes	Medium		Yes	Yes			
COFFEE1	Unknown	286	15	1			High		Yes	Yes			
COFFEE2	Unknown	286	15	1			High		Yes	Yes			
CBF1	Unknown	128	11	1	Yes	Yes	Low		Yes				Yes
CBF2	Unknown	128	13	1	Yes	Yes	Low		Yes				Yes
WORDS1	Text Recognition	2153	14	1	Yes	Yes	Medium		Yes			Yes	
WORDS2	Text Recognition	270	14	1	Yes	Yes	Medium	Yes	Yes			Yes	
WAFER1	Semiconductor	152	21	2	Yes	Yes	High	Yes	Yes				Yes
WAFER2	Semiconductor	152	84	4	Yes	Yes	High	Yes	Yes				Yes
ECG1	Medical	96	52	3			High	Yes	Yes				Yes
ECG2	Medical	96	55	2			High	Yes	Yes				Yes
FACEALL1	Face Recognition	131	74	2	Yes	Yes	Medium	Yes	Yes				Yes
FACEALL2	Face Recognition	131	140	2	Yes	Yes	Low	Yes	Yes				Yes

Table B.3 Summary of all time series sets (Cont)

Dataset	Domain	Length	N	O	Normal Series			Anomalous Series		Dissimilarity to normal series			
					Lagging	Amplitude	Similarity	Lagging	Amplitude	<10%	<30%	<60%	≥60%
*SBANK	Finance	426	12	1		Yes	Low		Yes				Yes
*SMINE	Finance	426	10	1			High		Yes				Yes
*YOGA1	Image	426	11	1	Yes		High		Yes			Yes	
YOGA2	Image	426	22	1	Yes		High		Yes		Yes		
*SWEDL1	Image	509	14	1	Yes		Medium		Yes				Yes
SWEDL2	Image	528	36	1	Yes		Medium		Yes		Yes		
FISH1	Image	463	22	1	Yes		High		Yes	Yes			
FISH2	Image	463	26	1	Yes		High		Yes		Yes		
*OLIVE1	Unknown	570	9	1			High		Yes	Yes			
*OLIVE2	Unknown	570	14	1			High		Yes	Yes			
TWOPT1	Synthetic	128	104	3	Yes	Yes	High	Yes	Yes	Yes			
TWOPT2	Synthetic	128	128	3		Yes	Medium		Yes	Yes			
*LIGHTING1	Synthetic	637	9	1	Yes	Yes	Low	Yes	Yes			Yes	
*LIGHTING2	Synthetic	637	12	1	Yes	Yes	Low	Yes	Yes			Yes	
*FACE41	Face Recognition	350	14	1	Yes		High	Yes	Yes				Yes
*FACE42	Face Recognition	350	14	1	Yes		High	Yes	Yes				Yes
GUNPT1	Video surveillance	150	76	1	Yes	Yes	High		Yes	Yes			
GUNPT2	Video surveillance	150	76	1	Yes	Yes	High		Yes	Yes			
GUNPT3	Video surveillance	150	76	1	Yes	Yes	High		Yes	Yes			

Time Lagging   Amplitude Differences

Fig. B.1 Typical time series problems—Two series are $x_a(t)$ and $x_b(t)$. Time lagging: $x_a(t) = x_b(t + x)$; Amplitude differences: $x_a(t) <> x_b(t)$

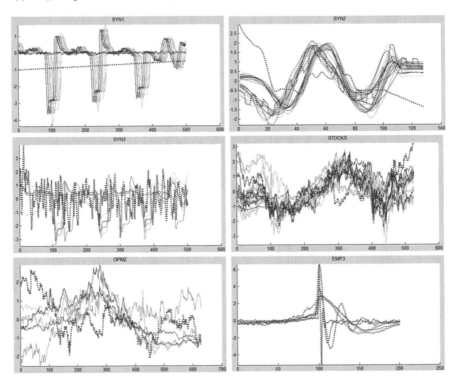

Fig. B.2 Plots of first six time series dataset of Table B.2. *Red dotted line* represents anomalous series

of zero and standard deviation of one. Since Wal-Mart is not a commodity, the outlier detection method is expected to find Wal-Mart as an outlier in this experiment.

- **Synthetic Lightning EMP (EMP3).** This data set, from [64], contains 11 time series; 8 of them are from one class and three anomalous series are from another class. Each series contains 201 observations.

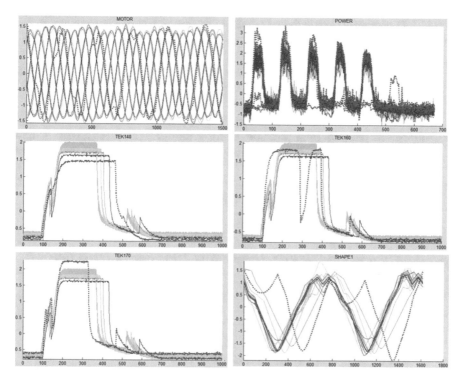

Fig. B.3 Plots of time series 7 to 12 of the dataset of Table B.2. *Red dotted line* represents anomalous series

- **Motor Current data set (MOTOR).** Original data set is from [69], and contains 420 time series. 21 were chosen including 1 anomalous time series, and each one consists of 1500 real values. Normal time series are the current signals measured from normal operations of a induction motor. The anomalous time series is obtained from a faulty motor.
- **Power Usage Data (POWER).** This data set was obtained from UCR [69], and contains 51 time series corresponding to the weekly power consumption measured every 15 min at a research facility from week 2 to week 52 in 1977. Each time series contains 672 values, and the anomalous time series represent the power consumption during the weeks with a holiday or special event.
- **NASA Valve Data (TEK140, TEK160 and TEK170).** This data set was also obtained from UCR [69]. The data values are solenoid current measurements on a Marotta MPV-41 series valve which is on and off under various test conditions in a laboratory. The normal time series correspond to the data measured during the normal operations of the valves; the time series data measured during a faulty operation of the valve is considered an anomaly. Three data sets T14, T16, and

T17 all contain 5 time series of which 4 are normal and one is anomalous. The anomalous time series were measured during different faulty operations of the valves.

- **Shape (SHAPE1 and SHAPE2)** These data sets were also obtained from UCR [69]. Both consist of 21 time series corresponding to the shapes. The normal time series have the shapes of bones while the anomalous time series has the shape of a cup.
- **Automatic Diatom Identification using Contour Analysis (ADIAC1 and ADIAC2)**. These data sets ,also obtained from UCR [69], describe the contours of different type of diatoms.
- **Words (WORDS1 and WORDS2)**. These data were also obtained from UCR [69] and describe the contours of word images.

B.1.3   Datasets for Online Anomalous Time Series Detection

To assess the effectiveness of the proposed algorithm we use eleven data sets, consisting of three synthetic data sets and eight modified real data sets introduced in the previous section, viz., SYN1, SYN2, SYN3, STOCKS, OPMZ, MOTOR, POWER, TEK140, TEK160, TEK170 and SHAPE1.

The data series in data sets are plotted in Fig. B.4.

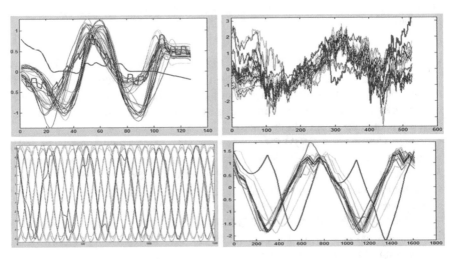

Fig. B.4 Four datasets used in the experiments (*from top to bottom, left to right*) are: SYN2, STOCKS, MOTOR, and SHAPE1. Anomalous series are marked as *red*

Table B.4 Time series data sets details

Dataset	Source	Length	w	Domain
SYN0	Generated	966	40	Synthetic
ECG1	[115, 116]	1000	40	Medical
ECG2	[115, 116]	2160	40	Medical
TEK140	[25]	5000	100	Electronic
TEK160	[25]	5000	100	Electronic
TEK170	[25]	5000	100	Electronic
VIDEOS1	[25]	11251	200	Video surveillance
VIDEOS2	[25]	11251	200	Video surveillance

B.1.4   Data Sets for Abnormal Subsequence Detection in a Single Series

We use one synthetic data set and 7 real data sets. The synthetic data set contains several copies of 'normal' data and two abnormal subsequences, both exactly the same. Real data sets are obtained from multiple application domains: two data sets are from medical domain, three are from electronic domain and two are from video surveillance domain. The details of data sets are given in Table B.4. The length of w for each data set is chosen either by following the original author's suggestion or by our observations.

B.1.5   Results for Abnormal Subsequence Detection in a Single Series for Various Datasets

Anomaly detection of a subsequence, obtained for datasets described in the previous section is shown in the following figures (Figs. B.5, B.6, B.7, B.8, B.9, B.10, B.11).

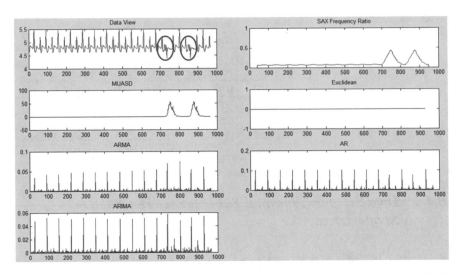

Fig. B.5 Experimental results for SYN0. *Red circles* highlight abnormal subsequences. (*Top Left*) Plot of SYN0 time series; (Other) Results of 6 algorithms used in these comparisons. Y-axis represents anomaly scores at time t. X-axis shows time t

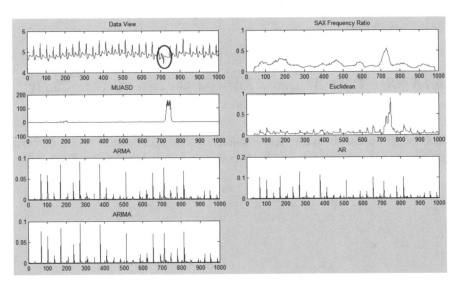

Fig. B.6 Experimental results for ECG1. *Red circles* highlight abnormal subsequences. (*Top Left*) Plot of ECG1 time series; (Other) Results of 6 algorithms used in these comparisons. Y-axis represents anomaly scores at time t. X-axis shows time t

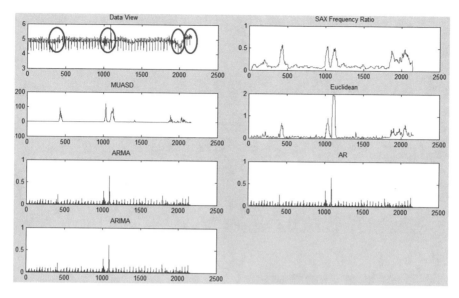

Fig. B.7 Experimental results for ECG2. *Red circles* highlight abnormal subsequences. (*Top Left*) Plot of ECG2 time series; (Other) Results of 6 algorithms used in these comparisons. Y-axis represents anomaly scores at time *t*. X-axis shows time *t*

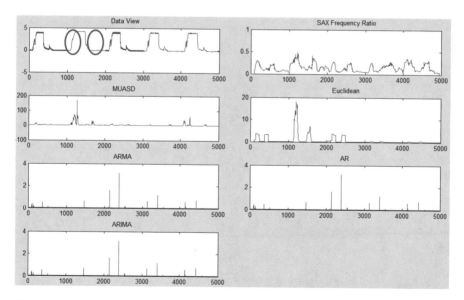

Fig. B.8 Experimental results for ECG2. *Red circles* highlight abnormal subsequences. (*Top Left*) Plot of ECG2 time series; (Other) Results of 6 algorithms used in these comparisons. Y-axis represents anomaly scores at time *t*. X-axis shows time *t*

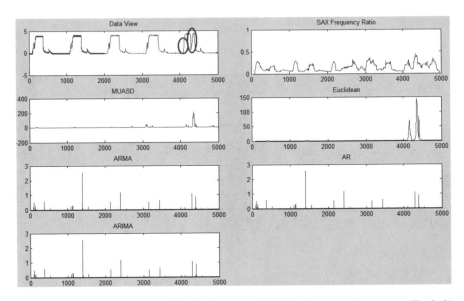

Fig. B.9 Experimental results for TK160. *Red circles* highlight abnormal subsequences. (*Top Left*) Plot of TK160 time series; (Other) Results of 6 algorithms used in these comparisons. Y-axis represents anomaly scores at time t. X-axis shows time t

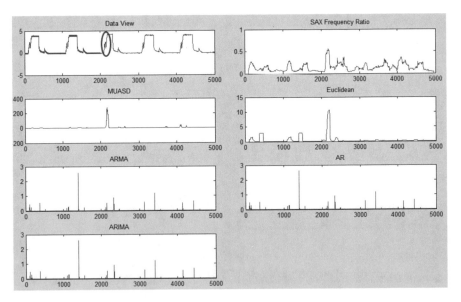

Fig. B.10 Experimental results for TK170. *Red circles* highlight abnormal subsequences. (*Top Left*) Plot of TK170 time series; (Other) Results of 6 algorithms used in these comparisons. Y-axis represents anomaly scores at time t. X-axis shows time t

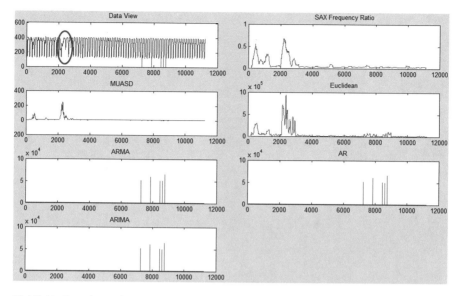

Fig. B.11 Experimental results for VIDEOS2. *Red circle* highlights abnormal subsequences. (*Top Left*) Plot of VIDEOS2 time series; (*Other*) Results of 6 algorithms used in these comparisons. Y-axis represents anomaly scores at time t. X-axis shows time t

References

1. N. Abe, B. Zadrozny, J. Langford, "Outlier detection by active learning," in *Proceedings of the 12th ACM SIGKDD International Conference on Knowledge Discovery and Data Mining* (ACM, New York, 2006), pp. 504–509
2. J. Abfalg, H.-P. Kriegel, P. Kroger, P. Kunath, A. Pryakhin, M. Renz, "Similarity search on time series based on threshold queries," in *Proceeding EDBT'06 Proceedings of the 10th International Conference on Advances in Database Technology* (Springer, Berlin, Heidelberg, 2006), pp. 276–294
3. C. Aggarwal, "Outlier ensembles: position paper," *ACM SIGKDD Explorations Newsletter*, vol. 14, pp. 49–58, 2013. [Online]. Available: http://dl.acm.org/citation.cfm?id=2481252
4. C.C. Aggarwal, *Outlier Analysis* (Springer Science & Business Media, New York, 2013)
5. C.C. Aggarwal, C.K. Reddy, *Data Clustering: Algorithms and Applications* (CRC Press, Boca Raton, 2013)
6. H. Akaike, "A new look at the statistical model identification." IEEE Trans. Automatic Control **19**(6), 716–723 (1974)
7. F. Angiulli, C. Pizzuti, "Fast outlier detection in high dimensional spaces," in *Principles of Data Mining and Knowledge Discovery* (Springer, New York, 2002), pp. 15–27
8. D. Asteriou, S.G. Hall, *Applied Econometrics* (Palgrave Macmillan, New York, 2011)
9. M.J. Azur, E.A. Stuart, C. Frangakis, P.J. Leaf Multiple imputation by chained equations: what is it and how does it work?. Int. J. Methods Psychiatr. Res. **20**, 40–49 (2011)
10. R. Baeza-Yates, B. Ribeiro-Neto, *Modern Information Retrieval* (Addison-Wesley Longman Publishing, Boston, 1999)
11. D.J. Berndt, J. Clifford, "Using dynamic time warping to find patterns in time series," in *AAAI Working Notes of the Knowledge Discovery in Databases Workshop*, pp. 359–370, 1994
12. J.C. Bezdek, *Pattern Recognition with Fuzzy Objective Function Algorithms* (Kluwer Academic Publishers, Norwell, 1981)
13. G. Bonanno, F. Lillo, R.N. Mantegna, "High-frequency cross-correlation in a set of stocks." Quant. Finan. **1**, 96–104 (2001)
14. S. Boriah, V. Chandola, V. Kumar, "Similarity measures for categorical data: A comparative evaluation." Red **30**(2), 3 (2008)
15. G.E. Box, G.M. Jenkins, G.C. Reinsel, *Time Series Analysis: Forecasting and Control* (Wiley, New York, 2013)
16. M.M. Breunig, H.-P. Kriegel, R.T. Ng, J. Sander, "LOF: Identifying density-based local outliers," in *Proceedings of the ACM SIGMOD International Conference on Management of Data* (ACM, New York, 2000), pp. 93–104

© Springer International Publishing AG 2017
K.G. Mehrotra et al., *Anomaly Detection Principles and Algorithms*, Terrorism, Security, and Computation, https://doi.org/10.1007/978-3-319-67526-8

17. H. Cao, G. Si, Y. Zhang, L. Jia, "Enhancing effectiveness of density-based outlier mining scheme with density-similarity-neighbor-based outlier factor." Expert Syst. Appl. Intl. J. **37**(12), (2010)

18. G. Chaitin, "Epistemology as information theory: from leibniz to omega.' arXiv preprint math/0506552 (2005)

19. K.P. Chan, A.W.C. Fu, "Efficient time series matching by wavelets," in *Proceeding ICDE '99 Proceedings of the 15th International Conference on Data Engineering, Sydney, Austrialia, March 23–26, 1999*, p. 126, 1999

20. V. Chandola, D. Cheboli, V. Kumar, "Detecting anomalies in a time series database," *Department of Computer Science and Engineering, University of Minnesota*, Tech. Rep. TR pp. 09–004, 2009

21. V. Chandola, V. Mithal, V. Kumar, "A comparative evaluation of anomaly detection techniques for sequence data," in *2008. ICDM '08. Eighth IEEE International Conference on Data Mining*, pp. 743–748, 2008

22. W. Chauvenet, *A Manual of Spherical and Practical Astronomy, V. II. 1863. Reprint of 1891. 5th edn* (Dover, New York, NY, 1960)

23. D. Cheboli, "Anomaly detection of time series," Ph.D. dissertation, University of Minnesota, 2010

24. L. Chen, R. Ng, "On the marriage of lp-norms and edit distance," in *VLDB '04: Proceedings of the Thirtieth International Conference on Very Large Data Bases*, pp. 792–803, 2004

25. Y. Chen, E. Keogh, B. Hu, N. Begum, A. Bagnall, A. Mueen, G. Batista, "The ucr time series classification archive," July 2015, www.cs.ucr.edu/~eamonn/time_series_data/

26. Y. Chen, M.A. Nascimento, B. Chin, O. Anthony, K.H. Tung, "Spade: On shape-based pattern detection in streaming time series," in *Data Engineering, 2007. ICDE 2007. IEEE 23rd International Conference, The Marmara Hotel, Istanbul, Turkey, April 15–20, 2007*, pp. 786–795, 2007

27. P.L. Combettes, V.R. Wajs, "Signal recovery by proximal forward-backward splitting." Multiscale Model. Simul. **4**(4), 1168–1200 (2005)

28. M. de Condorcet, Essay on the Application of Analysis to the Probability of Majority Decisions (in French), (1785)

29. C. Cortes, V. Vapnik, "Support-vector networks." Mach. Learn. **20**(3), 273–297 (1995)

30. C. De Mol, E. De Vito, L. Rosasco, "Elastic-net regularization in learning theory." J. Complexity **25**(2), 201–230 (2009)

31. F. Dellaert, "The expectation maximization algorithm" (2002)

32. A.P. Dempster, N.M. Laird, D.B. Rubin, "Maximum likelihood from incomplete data via the em algorithm." J. R. Stat. Soc. Ser. B (Methodological), 1–38 (1977)

33. H. Ding, G. Trajcevski, P. Scheuermann, X. Wang, E. Keogh, "Querying and mining of time series data: experimental comparison of representations and distance measures." Proc. VLDB Endowment, 1542–1552 (2008)

34. L. Ertoz, M. Steinbach, V. Kumar, "A new shared nearest neighbor clustering algorithm and its applications," in *Workshop on Clustering High Dimensional Data and Its Applications at 2nd SIAM International Conference on Data Mining*, pp. 105–115, 2002

35. L. Ertöz, M. Steinbach, V. Kumar, "Finding clusters of different sizes, shapes, and densities in noisy, high dimensional data," in *SDM* (SIAM, 2003)

36. M. Ester, H.P. Kriegel, J. Sander, X. Xu, "A density-based algorithm for discovering clusters in large spatial databases with noise," in *In Proceedings of the 2nd International Conference on Knowledge Discovery and Data Mining* (AAAI Press, 1996), pp. 226–231

37. B. Everitt, S. Landau, M. Leese, D. Stahl, "Cluster Analysis (Wiley Series in Probability and Statistics)," 2011

38. A. Fabrizio, P. Clara, "Outlier mining in large high-dimensional data sets." IEEE Trans. Knowledge Data Eng. **17**(2), 203–215 (2005)

39. A. Fabrizio, B. Stefano, P. Clara, "Distance-based detection and prediction of outliers." IEEE Trans. Knowledge Data Eng. **18**(2), 145–160 (2006)

40. C. Faloutsos, M. Ranganathan, Y. Manolopoulos., "Fast subsequence matching in time-series databases," in *Proceedings of the 1994 ACM SIGMOD International Conference on Management of Data, New York, NY, USA*, pp. 419–429, 1994

41. H. Fan, O.R. Zaïane, A. Foss, J. Wu, "Resolution-based outlier factor: detecting the top-n most outlying data points in engineering data." Knowledge Inform. Syst. **19**(1), 31–51 (2009)

42. J. Feng, Y. Sui, C. Cao, "Some issues about outlier detection in rough set theory." Expert Syst. Appl. **36**(3), 4680–4687 (2009)

43. A.J. Fox, "Outliers in time series." J. R. Stat. Soc. Ser. B (Methodological) **34**(3), 350–363 (1972)

44. Y. Freund, R. Schapire, "A desicion-theoretic generalization of on-line learning and an application to boosting." Comput. Learn. Theory **55**, 119–139 (1995). [Online]. Available: http://link.springer.com/chapter/10.1007/3-540-59119-2_166

45. R. Fujimaki, T. Yairi, K. Machida, "An anomaly detection method for spacecraft using relevance vector," in *Learning, The Ninth Pacific-Asia Conference on Knowledge Discovery and Data Mining (PAKDD)* (Springer, New York, 2005), pp. 785–790

46. Fusion Media Limited, "Commodity prices," Website, 2012. http://www.forexpros.com/commodities/real-time-futures

47. Z. Gao, "Application of cluster-based local outlier factor algorithm in anti-money laundering," in *International Conference on Management and Service Science, 2009. MASS'09* (IEEE, Washington, DC, 2009), pp. 1–4

48. B. Gould, "On peirce's criterion for the rejection of doubtful observations, with tables for facilitating its application." Astronomical J. **IV**, 83 (1855)

49. F.E. Grubbs, "Procedures for detecting outlying observations in samples." Technometrics **11**(1), 1–21 (1969)

50. S. Guha, R. Rastogi, K. Shim, "Cure: An efficient clustering algorithm for large databases," in *Proceedings of the 1998 ACM SIGMOD International Conference on Management of Data, Seattle, Washington, USA*, pp. 73–84, 1998

51. S. Guha, R. Rastogi, K. Shim, "Rock: A robust clustering algorithm for categorical attributes," in *15th International Conference on Data Engineering, 1999. Proceedings*, pp. 512–521, 2000

52. M. Gupta, J. Gao, C.C. Aggarwal, J. Han, "Outlier detection for temporal data: A survey." IEEE Trans. Knowledge Data Eng. **26**(9), 2250–2267 (2014)

53. B. Grofman, G. Owen, S.L. Feld, Thirteen theorems in search of the truth. (PDF). Theory & Decision. **15**(3), 261–78 (1983)

54. J.D. Hamilton, *Time Series Analysis* (Princeton University Press, Princeton, 1994)

55. J. Han, M. Kamber, J. Pei, "Data Mining: Concepts and Techniques (The Morgan Kaufmann Series in Data Management Systems)," 2006

56. J. Hawthorne, "Voting in search of the public good: the probabilistic logic of majority judgements." Available at http://james-hawthorne.oucreate.com/ (2006)

57. A.E. Hoerl, R.W. Kennard, "Ridge regression: Biased estimation for nonorthogonal problems." Technometrics **12**(1), 55–67 (1970)

58. K. Hori, "Fault detection by mining association rules from housekeeping data," 2001

59. H. Huang, K. Mehrotra, C.K. Mohan, "Rank-based outlier detection." J. Stat. Comput. Simul. **82**, 1–14 (2011)

60. H. Huang, K. Mehrotra, C. Mohan, "Detection of anomalous time series based on multiple distance measures," in *28th International Conference on Computers and Their Applications (CATA-2013), Honolulu, Hawaii, USA*, 2013

61. C.M. Hurvich, C.-L. Tsai, "Regression and time series model selection in small samples." Biometrika **76**(2), 297–307 (1989)

62. A.K. Jain, R.C. Dubes, *Algorithms for Clustering Data* (Prentice-Hall, Upper Saddle River, 1988)

63. R.A. Jarvis, E.A. Patrick, "Clustering using a similarity measure based on shared near neighbors." IEEE Trans. Comput. **100**(11), 1025–1034 (1973)

64. C. Jeffery, "Synthetic lightning emp data," 2005. [Online]. Available: http://nis-www.lanl. gov/~eads/datasets/emp

65. W. Jin, A.K.H. Tung, J. Han, W. Wang, "Ranking outliers using symmetric neighborhood relationship," in *Pacific-Asia Conference on Knowledge Discovery and Data Mining*, pp. 577–593, 2006

66. S. Kaniovski, Z. Alexander, Optimal Jury Design for Homogeneous Juries with Correlated Votes. Theory and Decision **71**(4), 439–459 (2011)

67. E. Keogh, S. Lonardi, B. Chiu, "Finding surprising patterns in a time series database in linear time and space," in *Proc. of the 8th ACM SIGKDD International Conf on Knowledge Discovery and Data Mining*, pp. 550–556, 2002

68. E. Keogh, C.A. Ratanamahatana, "Exact indexing of dynamic time warping." Knowl. Inf. Syst. **3**, 358–386 (2005)

69. E. Keogh, Q. Zhu, B. Hu, Y. Hao, X. Xi, L. Wei, C. Ratanamahatana, "The UCR time series classification/clustering," Website, 2011. http://www.cs.ucr.edu/~eamonn/time_series_data/

70. E. Keogh, K. Chakrabarti, M. Pazzani, S. Mehrotra, "Dimensionality reduction for fast similarity search in large time series databases." J. Knowl. Inform. Syst., 263–283 (2000)

71. E. Keogh, S. Kasetty, "On the need for time series data mining benchmarks: A survey and empirical demonstration," in *In the 8th ACM SIGKDD International Conference on Knowledge Discovery and Data Mining*, pp. 102–111, 2002

72. E. Keogh, J. Lin, A. Fu, "Hot sax: Efficiently finding the most unusual time series subsequence," in *Fifth IEEE International Conference on Data Mining* (IEEE, Washington, DC, 2005), pp. 8–pp.

73. E. Keogh, J. Lin, S.-H. Lee, H.V. Herle, "Finding the most unusual time series subsequence: algorithms and applications." Knowl. Inform. Syst. **11**(1), 1–27 (2007)

74. E. Keogh, S. Lonardi, C.A. Ratanamahatana, "Towards parameter-free data mining," in *Proceedings of the Tenth ACM SIGKDD International Conference on Knowledge Discovery and Data Mining* (ACM, New York, 2004), pp. 206–215

75. E.M. Knorr, R.T. Ng, "A unified notion of outliers: Properties and computation," in *KDD*, pp. 219–222, 1997

76. E.M. Knorr, R.T. Ng, "Algorithms for mining distance-based outliers in large datasets," in *Proceeding VLDB '98 Proceedings of the 24rd International Conference on Very Large Data Bases*, pp. 392–403, 1998

77. T. Kohonen, "The self-organizing map." Proc. IEEE **78**(9), 1464–1480 (1990)

78. T.S. Kuhn, *The Structure of Scientific Revolutions, 2nd enl. edn* (University of Chicago Press, Chicago, 1970)

79. J. Lin, E. Keogh, L. Wei, S. Lonardi, "Experiencing sax: a novel symbolic representation of time series." *Data Mining and Knowledge Discovery*, pp. 107–144, 2007

80. J. Lin, R. Khade, Y. Li, "Rotation-invariant similarity in time series using bag-of-patterns representation." J. Intell. Inform. Syst. (1 April 2012), 1–29 (2012)

81. S. Lloyd, "Least squares quantization in pcm." IEEE Trans. Inform. Theory **28**(2), 129–137 (1982)

82. J. Ma, S. Perkins, A.B. Corp, "Online novelty detection on temporal sequences," pp. 613–618, 2003

83. J. Ma, J. Theiler, S. Perkins, "Accurate on-line support vector regression." Neural Comput. **15**(11), 2683–2703 (2003)

84. P.C. Mahalanobis, "On the generalized distance in statistics," in *Proceedings of the National Institute of Sciences (Calcutta)*, vol. 2, pp. 49–55, 1936

85. K. Mehrotra, C. Mohan, S. Ranka, *Elements of Artificial Neural Networks* (MIT Press, Cambridge, MA, 1997)

86. X. Meng, Z. Chen, "On user-oriented measurements of effectiveness of web information retrieval systems," in *Proceeding of the 2004 International Conference on Internet Computing*, pp. 527–533, 2004

87. H.Z. Moayedi, M. Masnadi-Shirazi, "Arima model for network traffic prediction and anomaly detection," in *ISIT*, vol. 4, pp. 1–6, 2008

88. H.Z. Moayedi, M. Masnadi-Shirazi, , "Arima model for network traffic prediction and anomaly detection," in *International Symposium on Information Technology, 2008. ITSim 2008*, vol. 4 (IEEE, Washington, DC, 2008), pp. 1–6

89. S. Morgan, "Cyber crime costs projected to reach \$2 trillion by 2019," 2016. [Online]. Available: http://www.forbes.com

90. M. Müller, "Dynamic time warping." *Information Retrieval for Music and Motion*, pp. 69–84, 2007

91. E.A. Nadaraya, "On estimating regression." Theory Probab. Appl. **9**(1), 141–142 (1964)

92. S. Papadimitriou, H. Kitagawa, P.B. Gibbons, C. Faloutsos, "Loci: Fast outlier detection using the local correlation integral," in *Proceedings. 19th International Conference on Data Engineering, 2003* (IEEE, Washington, DC, 2003), pp. 315–326

93. B. Peirce, "Criterion for the rejection of doubtful observations." Astron. J. II **45** (1852)

94. B. Pincombe, "Anomaly detection in time series of graphs using arma processes." ASOR Bull. **24**(4), 2–10 (2005)

95. B. Pincombe, "Anomaly detection in time series of graphs using arma processes." Asor Bull. **24**(4), p. 2 (2005)

96. P. Protopapas, J. M. Giammarco, L. Faccioli, M.F. Struble, R. Dave, C. Alcock, "Finding outlier light curves in catalogues of periodic variable stars." Mon. Not. R. Astron. Soc. **369**(2), 677–696 (2006)

97. T. Rakthanmanon, B. Campana, A. Mueen, G. Batista, B. Westover, Q. Zhu, J. Zakaria, E. Keogh, "Searching and mining trillions of time series subsequences under dynamic time warping," in *Proceeding KDD '12 Proceedings of the 18th ACM SIGKDD International Conference on Knowledge Discovery and Data Mining*, pp. 262–270, 2012

98. S. Ramaswamy, R. Rastogi, K. Shim, "Efficient algorithms for mining outliers from large data sets," in *ACM SIGMOD Record*, vol. 29(2) (ACM, New York, 2000), pp. 427–438

99. C.A. Ratanamahatana, E. Keogh, "Three myths about dynamic time warping," in *Proceedings of SIAM International Conference on Data Mining (SDM '05), Newport Beach, CA, April 21–23*, pp. 506–510, 2005

100. J. Rissanen, "Modeling by shortest data description." Automatica 14(5), 465–471 (1978)

101. P.J. Rousseeuw, A.M. Leroy, *Robust Regression and Outlier Detection* (Wiley, New York, 1987)

102. M.H. Safar, C. Shahabi, "Multidimensional index structures." *Shape Analysis and Retrieval of Multimedia Objects* (Springer, New York, 2003), pp. 63–77

103. G. Salton, *Automated Text Processing: The Transformation, Analysis, and Retrieval of Information by Computer* (Addison-Wesley Longman Publishing, Boston, 1998)

104. R. Sibson, "Slink: an optimally efficient algorithm for the single-link cluster method." Comput. J. **16**(1), 30–34 (1973)

105. A. Smola, V. Vapnik, "Support vector regression machines." Adv. Neural Inform. Process. Syst. **9**, 155–161 (1997)

106. A.J. Smola, B. Schölkopf, "A tutorial on support vector regression." Stat. Comput. **14**(3), 199–222 (2004)

107. J. Tang, Z. Chen, A.W. Fu, D.W. Cheung, "Capabilities of outlier detection schemes in large datasets, framework and methodologies." Knowl. Inform. Syst. **11**(1), 45–84 (2006)

108. J. Tang, Z. Chen, A.W. chee Fu, D.W. Cheung, "Enhancing effectiveness of outlier detections for low density patterns," in *Proceedings of the Pacific-Asia Conference on Knowledge Discovery and Data Mining*, pp. 535–548, 2002

109. Y. Tao, D. Pi, "Unifying density-based clustering and outlier detection," in *2009 Second International Workshop on Knowledge Discovery and Data Mining, Paris, France*, pp. 644–647, 2009

110. R. Tibshirani, "Regression shrinkage and selection via the lasso." J. R. Stat. Soc. Ser. B (Methodological) 267–288 (1996)

111. A.N. Tikhonov, V.Y. Arsenin, "Solutions of ill-posed problems," 1977

112. V. Vapnik, *Estimation of Dependences Based on Empirical Data* (Springer Science & Business Media, New York, 2006)

113. M. Vlachos, D. Gunopoulos, G. Kollios, "Discovering similar multidimensional trajectories," in *Proceeding ICDE '02 Proceedings of the 18th International Conference on Data Engineering, IEEE Computer Society Washington, DC, USA*, p. 673, 2002

114. G.S. Watson, "Smooth regression analysis." Sankhyā Indian J. Stat. Ser. A 359–372 (1964)

115. L. Wei, N. Kumar, V. Lolla, E. Keogh, S. Lonardi, C.A. Ratanamahatana, "Assumption-free anomaly detection in time series," in *Proceedings of the 17th International Conference on Scientific and Statistical Database Management (SSDBM'05)*, pp. 237–242, 2005

116. L. Wei, N. Kumar, V. Lolla, E. Keogh, S. Lonardi, C.A. Ratanamahatana, "Assumption-free anomaly detection in time series," Website, 2005. http://alumni.cs.ucr.edu/~wli/SSDBM05/

117. P. Werbos, "Beyond regression: New tools for prediction and analysis in the behavioral sciences," 1974

118. Yahoo Inc, "Stocks price dataset," Website, 2012. http://finance.yahoo.com/q/hp?s=WMT+Historical+Prices

119. B.-K. Yi, C. Faloutsos, "Fast time sequence indexing for arbitrary lp norms." VLDB (2000)

120. T. Zhang, R. Ramakrishnan, M. Livny, "Birch: an efficient data clustering method for very large databases," in *ACM Sigmod Record*, vol. 25(2) (ACM, New York, 1996), pp. 103–114

121. Y. Zhang, S. Yang, Y. Wang, "LDBOD: A novel local distribution based outlier detector." Pattern Recognit. Lett. **29**(7), 967–976 (2008)

122. Z. Zhao, K.G. Mehrotra, C.K. Mohan, "Ensemble algorithms for unsupervised anomaly detection," in *International Conference on Industrial, Engineering and Other Applications of Applied Intelligent Systems* (Springer, New York, 2015), pp. 514–525

123. Z. Zhao, C.K. Mohan, K. Mehrotra, "Adaptive sampling and learning for unsupervised outlier detection," in *The 29th International Florida Artificial Intelligence Research Society*, 2016

124. J. Zhou, Y. Fu, Y. Wu, H. Xia, Y. Fang, H. Lu, "Anomaly detection over concept drifting data streams." J. Comput. Inform. Syst. **5**, 1697–1703 (2009)

125. A. Zimek, R.J. Campello, J. Sander, "Ensembles for unsupervised outlier detection: challenges and research questions a position paper." ACM SIGKDD Explorations Newsletter **15**(1), 11–22 (2014)

126. A. Zimek, M. Gaudet, R.J. Campello, J. Sander, "Subsampling for efficient and effective unsupervised outlier detection ensembles," in *Proceedings of the 19th ACM SIGKDD International Conference on Knowledge Discovery and Data Mining* (ACM, New York, 2013), pp. 428–436

Index

Symbols
k-means clustering, 41, 43

A
Abnormal subsequence detection, 167
Active learning, 135, 142
Active-outlier, 149
Adaboost, 135, 141
Adaptive learning, 142
Adaptive sampling, 135
Agglomerative Clustering, 46
Aggregation approach, 158
Akaike Information Criterion, 31, 44
Anomaly in normal distribution, 97
ARIMA, 68, 74
ARMA, 68
Auto-Regression, 68

B
Bankruptcy prediction, 10
Battlefield Behaviors, 13
Bayesian Information Criteria, 45
Between time series detection, 155
BIRCH, 46

C
Change detection, 79
Cluster density effect, 127
Cluster-based LOF, 111
Clustering, 27
COF, 140
Contextual anomaly, 156

Correlations with delays, 90
Cosine similarity measure, 34
Credit card fraud, 9
Creditworthiness, 10
CURE, 46, 48
Customer behavior, 17
Cybersecurity, 7

D
Data space, 57, 59
DB outlier, 100
DBSCAN, 47–49
Density based approach, 107
Density factor, 126
Density-based clustering, 126
Diagnosis prediction, 11
DIFFSTD, 169
Discords, 156
Discrete Fourier Transform, 76, 91, 163
Discrete wavelet transform, 91, 163
Discretization approach, 158
Distance-based Anomaly Detection, 29
Divisive Clustering, 48
Dynamic time warping (DTW), 163

E
EDR, 162
Elastic measures, 162
elbow method, 44
Employee behavior, 17
Ensemble methods, 135
Epidemiology, 12
Explicit models, 61

K.G. Mehrotra et al., *Anomaly Detection Principles and Algorithms*, Terrorism, Security, and Computation, https://doi.org/10.1007/978-3-319-67526-8

F
Fraudulent email, 8
Fuzzy Clustering, 45

G
Gauss-Newton, 65

H
Haar Transformation, 76
Haar wavelets, 76

I
Implicit model, 59
Independent ensemble, 135
Index-based algorithm, 100
Influential measure of outlierness, 114
Information theory, 31
Information-theoretic Criteria, 44
Investing, 10

J
Jaccord measure, 34

K
Kernel based density, 130
Kernel regression, 31, 66
Kernel-based techniques, 69

L
Learning algorithms, 91
Linear regression, 65
Local correlation integral, 102
Local density, 26
Local outlier factor, 110
Lock-step measures, 160
Longest common subsequence (LCSS), 163

M
Mahalanobis distance, 26, 34
Majority voting, 138
Malware detection, 8
malware infection, 25
Manhattan distance, 35
Markov models, 70
Markovian techniques, 69
Marquardt parameter, 65
Maximum likelihood approach, 109

Min-rank method, 137
Minimum margin, 142
Minkowski Distance, 34
Mixture density estimation, 109
Mixture of Densities, 31
Model parameter approach, 58
Model-based methods, 68
Modified rank, 124
Multi-granularity deviation Factor, 102
Multimodal Distributions, 27

N
Nearest neighbor approach, 105
Nearest Neighbor Clustering, 42
Nearest neighbors, 37
Neigborhood clustering, 124
Neighborhood-based local density, 126
Neural networks, 24, 65
Nonlinear regression, 65

O
Occam's razor, 92
Online anomaly detection, 157
Online identification of anomalous series, 172
Outlier detection based on multiple measures,
      169

P
Parameter space, 57
Patient monitoring, 12
Personnel Behaviors, 13
Piecewise Aggregate Approximation (PAA),
      164
Point-to-point distance, 87

Q
Quality control, 16

R
RADA, 140
Radial Basis Function, 31
Radiology, 12
Rank based clustering, 124
Rank with averaged distance, 128
Rank-based algorithm, 29, 119, 121
Rate anomaly, 156
Regression, 68
Regression model, 64
Regularization, 31, 92

Resolution-based outlier, 104
Reverse nearest neighbors, 114

S
Sampling, 139
Sampling neighborhood, 102
SAX with bag-of-pattern (SAXBAG), 163, 166
Self Organizing Map, 55
semi-supervised learning, 30
Sequential approach, 139
Shared Nearest Neighbor, 34
Signal processing approach, 158
Similarity measures, 34
Spatial Assembling Distance SpADe, 163
Splines, 67
STREND, 169
supervised learning, 30
Support Vector Machines, 24, 31, 66
Support Vector Regression, 68

Symbolic Aggregate approXimation (SAX), 163

T
Time series models, 72
Time-varying processes, 67
TQuEST, 163

U
Unimodal Distributions, 27
unsupervised learning, 30

W
Weak learner, 135
Weighted adaptive sampling, 143
Window-based techniques, 69

Printed in the United States
By Bookmasters